Radical Design and Concrete Practices

Proceedings of the International Seminar
held at the University of Dundee, Scotland, UK
on 7 September 1999

Edited by

Ravindra K. Dhir
Director, Concrete Technology Unit
University of Dundee

and

Kevin A. Paine
Research/Teaching Fellow, Concrete Technology Unit
University of Dundee

Thomas Telford

Published by Thomas Telford Publishing, Thomas Telford Limited, 1 Heron Quay, London E14 4JD.

URL: http://www.t-telford.co.uk

Distributors for Thomas Telford books are
USA: ASCE Press, 1801 Alexander Bell Drive, Reston, VA 20191-4400, USA
Japan: Maruzen Co. Ltd, Book Department, 3–10 Nihonbashi 2-chome, Chuo-ku, Tokyo 103
Australia: DA Books and Journals, 648 Whitehorse Road, Mitcham 3132, Victoria

First published 1999

The full list of titles from the 1999 International Congress 'Creating with Concrete' and available from Thomas Telford is as follows

- *Creating with concrete*
- *Radical design and concrete practices*
- *Role of interfaces in concrete*
- *Controlling concrete degradation*
- *Extending performance of concrete structures*
- *Exploiting wastes in concrete*
- *Modern concrete materials: binders, additions and admixtures*
- *Utilizing ready-mixed concrete and mortar*
- *Innovation in concrete structures: design and construction*
- *Specialist techniques and materials in concrete construction*
- *Concrete durability and repair technology*

A catalogue record for this book is available from the British Library

ISBN: 0 7277 2817 2

Printed and bound in Great Britain by MPG Books, Bodmin, Cornwall

PREFACE

Concrete is the key material for Mankind to create the built environment, the requirements for which are both demanding in terms of technical performance and economy and yet greatly varied from architectural masterpieces to the simplest of utilities. This presents the greatest challenge and the question is how best to advance concrete and create imaginatively.

In response, the Concrete Technology Unit (CTU) of the University of Dundee organised this Congress following on from its established series of events, namely, Concrete in the Service of Mankind in 1996, Concrete 2000: Economic and Durable Concrete Construction Through Excellence in 1993 and Protection of Concrete in 1990.

Under the theme of Creating with Concrete, the Congress consisted of five Seminars: (i) Radical Design and Concrete Practices, (ii) Role of Interfaces in Concrete, (iii) Controlling Concrete Degradation, (iv) Extending Performance of Concrete Structures and (v) Exploiting Wastes in Concrete, and five Conferences: (i) Modern Concrete Materials: Binders, Additions and Admixtures, (ii) Utilising Ready-Mixed Concrete and Mortar, (iii) Innovation in Concrete Structures: Design and Construction, (iv) Specialist Techniques and Materials for Concrete and Construction and (v) Concrete Durability and Repair Technology. In all, a total of 421 papers were presented from 67 countries.

The Opening Addresses were given by Mr Henry McLeish, MP, MSP, Minister for Enterprise and Lifelong Learning, Scotland, Dr Ian Graham-Bryce, Principal and Vice Chancellor of Dundee University, Mrs Helen Wright, Lord Provost, City of Dundee and Professor Peter Hewlett, President of the Concrete Society. This was followed by four Opening Papers by leading international experts; Dr Bryant Mather, US Army Corps of Engineers, Professor Charles F Hendriks, Delft University of Technology, Netherlands, Dr Bjørn Jensen, Danish Technological Institute, Dr Oliver Kornadt, Philipp Holzmann AG, Germany, Professor Jurek Tolloczko, Concrete Society, UK, Mr Michael Téménidès, CIMBÉTON, France and Professor Yves Malier, Ecole Normale Superieure de Cachan, France. The Closing Address was given by Professor John Morris, University of the Witwatersrand, South Africa.

The support of 20 International Professional Institutions and 31 sponsors was a major contribution to the success of the Congress. An extensive Trade Fair, participated in by 50 organisations, formed an integral part of the Congress. The work of the Congress was an immense undertaking and all of those involved are gratefully acknowledged, in particular, the members of the Organising Committee for managing the event from start to finish; members of the International Advisory and National Technical Committees for advising on the selection and reviewing of papers; the Authors and the Chairmen of Technical Sessions for their invaluable contributions to the proceedings.

All of the proceedings have been prepared directly from the camera-ready manuscripts submitted by the authors and editing has been restricted to minor changes where it was considered absolutely necessary.

Dundee
September 1999

Ravindra K Dhir
Chairman, Congress Organising Committee

INTRODUCTION

Concrete is one of the most common and versatile of construction materials and forms the basis of most modern infrastructure. In its simplest form, it is a well established technology but one which has not been without its problems in terms of both image and performance. However, technology for concrete has not stood still and some of the more radical developments in design and construction can now be transferred to mainstream concrete practice.

The acceptance of radical new techniques and methods for using concrete is not without risk, and the construction industry does not necessarily facilitate innovation by an over-dependence on codes of practice and design specifications. Many engineers, therefore, tend not to adopt innovations until they become accepted techniques. This is understandable and it is necessary to provide suitable methods of technology transfer for new principles so that engineers have more confidence in their use. However, it is also necessary that codes of practice are not regulatory and allow a degree of freedom so that innovative techniques can be adopted without risk to the client. This is particularly pertinent to Europe, where EuroCodes and European Standards are being developed and introduced.

In addition to new principles, there is a need to take on board new practices, more readily, for concrete to compete economically with other construction materials. In this respect, precast concrete and composite construction are two advanced forms of structural technology which are too often overlooked as engineers use what they know. This is partly because the benefits of new practice are difficult to appreciate in the short term. Methods used to improve long-term appearance, energy-efficiency and durability are often difficult to quantify in fiscal terms. Thus, it can be difficult to justify the adoption of new practices unless they impact on tender prices.

As with any new technology there is a need for radical principles and practices in the concrete industry to be appropriately and correctly exploited. In this regard, it is essential that they are introduced with appropriate levels of education and training.

The Proceedings of this Seminar; *'Radical Design and Concrete Practices'* dealt with these subject areas and the issues raised, under two clearly identified themes: (i) New Principles and (ii) New Practices. The Proceedings opened with a Leader Paper presented by a foremost exponent in the field. There were a total of 21 papers presented during the one day International Seminar.

Dundee
May 1999

Ravindra K Dhir
Kevin A Paine

ORGANISING COMMITTEE
Concrete Technology Unit

Professor R K Dhir, OBE (Chairman)

Dr M R Jones (Secretary)

Mr M D Newlands (Joint Secretary)

Professor P C Hewlett
British Board of Agrément

Dr N A Henderson
Mott MacDonald Ltd

Professor V K Rigopoulou
National Technical University of Athens, Greece

Dr S Y N Chan
Hong Kong Polytechnic University

Dr N Y Ho
L & M Structural Systems, Singapore

Dr M J McCarthy

Dr M C Limbachiya

Dr T D Dyer

Dr K A Paine

Dr T G Jappy

Mr P A J Tittle

Mr J C Knights

Mr S R Scott (Unit Assistant)

Miss A M Duncan (Unit Secretary)

INTERNATIONAL ADVISORY COMMITTEE

INTERNATIONAL ADVISORY COMMITTEE (CONTINUED)

NATIONAL TECHNICAL COMMITTEE

Mr P Barber
Manager of the Scheme, The Quality Scheme for Ready Mixed Concrete

Professor A W Beeby
Professor of Structural Design, University of Leeds

Mr B V Brown
Divisional Technical Executive, Readymix (UK) Ltd.

Dr T W Broyd
Technology Development Director, W S Atkins Ltd.

Professor J H Bungey
Professor of Civil Engineering, University of Liverpool

Dr P S Chana
Director, CRIC, Imperial College of Science, Technology & Medicine

Professor J L Clarke
Principal Engineer, The Concrete Society

Dr P C Das
Group Manager, Structures Management, Highways Agency

Dr S B Desai, OBE
Principal Civil Engineer, Department of the Environment, Transport and the Regions

Professor R K Dhir, OBE (Chairman)
Director, Concrete Technology Unit, University of Dundee

Mr C R Ecob
Director Special Services Division, Mott MacDonald Ltd.

Professor F P Glasser
University of Aberdeen

Professor T A Harrison
Technical Consultant, Quarry Products Association

Professor P C Hewlett
Director, British Board of Agrément

Professor J Innes
Director of Roads, Scottish Office

NATIONAL TECHNICAL COMMITTEE (CONTINUED)

Mr K A L Johnson
Director, AMEC Civil Engineering Ltd.

Dr M R Jones
Senior Lecturer, Concrete Technology Unit, University of Dundee

Mr P Livesey
National Technical Services Manager, Castle Cement Ltd.

Professor A E Long
Director of School, Queens University of Belfast

Professor P S Mangat
Head of Research, Sheffield Hallam University

Mr G Masterton
Director, Babtie Group Ltd.

Professor G C Mays
Director of Civil Engineering, Cranfield University

Mr L H McCurrich
Technology Development Consultant, Fosroc Construction

Professor R S Narayanan
Partner, SB Tietz & Partners Consulting Engineers

Dr P J Nixon
Head, Centre for Concrete Construction, Building Research Establishment Ltd.

Dr W F Price
Senior Associate, Messrs Sandberg

Professor G Somerville, OBE
Director of Engineering, British Cement Association

Professor D C Spooner
Director, Materials and Standards, British Cement Association

Dr H P J Taylor
Director, Tarmac Precast Concrete Ltd.

Mr M Walker
Technical Manager, The Concrete Society

Dr R J Woodward
Senior Project Manager, Transport Research Laboratory

SUPPORTING INSTITUTIONS

American Concrete Institute, USA

American Society of Civil Engineers, USA

Australian Concrete Institute

Concrete Association of Finland

Concrete Society of Southern Africa

Concrete Society, UK

Danish Concrete Society, Denmark

Fédération de l'Industrie du Beton, France

German Concrete Association (DBV)

Hong Kong Institution of Engineers

Indian Concrete Institute

Institute of Concrete Technology, UK

Institution of Civil Engineers, UK

Instituto Brasileiro Do Concreto, Brazil

Japan Concrete Institute

Netherlands Concrete Society

New Zealand Concrete Society

Norwegian Concrete Association

Singapore Concrete Institute

Spanish Association for Structural Concrete

Swedish Concrete Association

SPONSORING ORGANISATIONS WITH EXHIBITION

AMEC Civil Engineering Ltd.

Babtie Group Ltd.

Bardon Aggregates

Blue Circle Cement

Blyth & Blyth

British Board of Agrément

British Cement Association

Building Research Establishment

Castle Cement Ltd.

Cementitious Slag Makers Association

CIMBÉTON, France

Du Pont de Nemours International S.A., Switzerland

ECC International Ltd.

Elkem Ltd. (Materials)

Fosroc International Ltd.

Grace Construction Products

HERACLES General Cement Co., Greece

John Doyle Group

Lafarge Aluminates

L M Scofield Europe Ltd.

Minelco Ltd.

Mott MacDonald Ltd.

O'Rourke Group

Ove Arup and Partners

SPONSORING ORGANISATIONS WITH EXHIBITION (CONTINUED)

Readymix (UK) Ltd.

Rugby Cement

Scottish Enterprise Tayside

Sika Ltd.

SKW - MBT Construction Chemicals

Thomas Telford Publishing Ltd.

United Kingdom Quality Ash Association

W A Fairhurst & Partners

ADDITIONAL EXHIBITORS

Christison Scientific Equipment Ltd.

CMS Pozament Limited

The Concrete Society

David Ball Group plc.

E & FN Spon

Flexcrete Ltd.

Germann Instruments A/S, Denmark

Natural Cement Distribution Limited

Palladian Publications Ltd.

Quality Scheme for Ready Mixed Concrete

UK Certification Authority for Reinforcing Steel

Wacker-Chemie GmbH, Germany

Wexham Developments

CONTENTS

LEADER PAPER

RADICAL DESIGN AND CONCRETE PRACTICE

R S Narayanan

S B Tietz and Partners

United Kingdom

ABSTRACT. Designing for sustainability will force all industries to re-think their current methods of design and construction. Life cycle analysis will be necessary to identify and minimise environmental impacts. Recycling materials and sites will also become the norm in the future. Designers can contribute by re-examining the design loadings and partial factors. The industry is taking steps to improve the quality of concrete construction. This in turn will also contribute to sustainability.

Keywords: Sustainable construction, Environmental impact, Life cycle analysis, Re-cycling, Design loads, Partial factors, High strength quality control issues, Whole life Costs, Certification, Operative training.

Professor R S "Nary" Narayanan is a Partner of S B Tietz & Partners, Consulting Engineers, London. He is a Visiting Professor at The University of Leeds in Principles of Design. He sits on a number of Technical Committees and chairs some of them. He is currently Chairman of The Concrete Industry Alliance Steering Group for Concrete 2005, the Innovation and Research Strategy.

INTRODUCTION

Concrete is an exceptionally versatile construction medium. It has found application in every major form of construction. It combines strength with durability. It can be moulded into shapes to produce aesthetically attractive structures. Many elegant, delicate and monumental concrete structures all around the world bear testament to the qualities of concrete. Concrete can be made with different properties to suit particular applications (e g) densities, strengths, pore structures. It can be pre-cast or placed in-situ. It can be pumped sprayed and placed under water.

In its long history, concrete industry has not been without its problems. This is not uncommon with any technological development. The press, with its love of "bad news", has helped to create a poor image for concrete from time to time. The media are not interested to note that the majority of concrete construction performs well. Responsible sections of the industry address the problems that manifest and analyse their causes. Where new measures are warranted, these are put into practice. It is certain that the industry will face challenges in the future and it will need to be constantly vigilant and be prepared to adapt.

A challenge, which is at our doorstep now, arises from society's awareness and concerns regarding environmental matters. A welter of legislation already exists in this area. I am sure that more is in the pipeline. The UK government is a signatory to a number of international treaties in this field and has committed itself to a number of environmental targets. Sustainable development is now more than just a slogan. This has enormous implications to the construction industry.

All sectors, be they material producers, designers, contractors or client bodies, will need to take into account the requirements of this extra dimension of sustainability in their sphere of activity. A number of studies and research are already underway to understand the issues for concrete construction and to provide methodologies to facilitate decision making.

In this keynote paper, I propose to survey some of the initiatives the industry has undertaken with the hope of informing a wider audience

PROFILE OF THE INDUSTRY

In any broad discussions covering the whole of an industry, it will be useful to have a feel for some vital statistics. I have drawn heavily from Concrete 2005 (DOE's Innovation and Research Strategy for the Concrete Industry Sector [1]) with which I was closely connected. Figure 1 shows the complex web of sectors which make up the industry. Figure 2 shows an overview of the industry, represented by a cube, the three axes of which being markets, materials and processes. Decisions are made at each of the blocks, which affect the use and performance of concrete in respect of other blocks. Figure 3 provides a feel for the amount of materials used in concrete construction. Figure 4 is a pie chart of the industry by types of construction

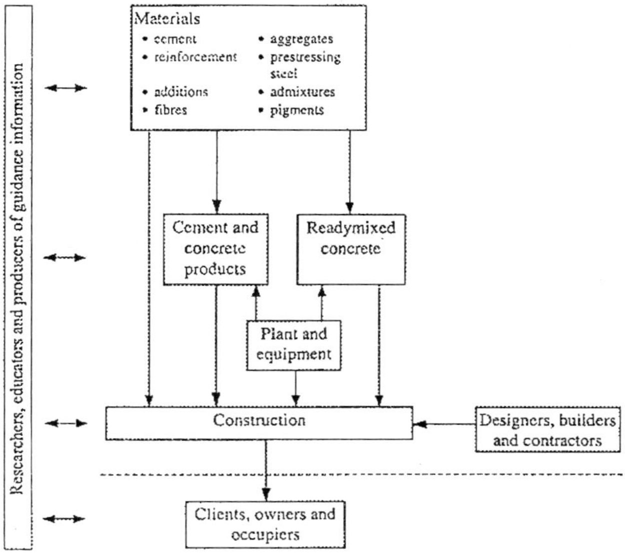

Figure 1 Definition of concrete industry sectors

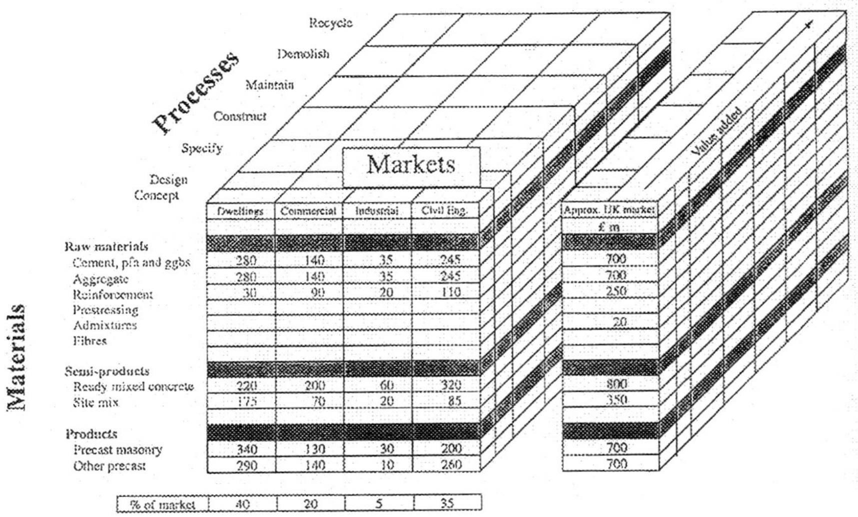

	Dwellings	Commercial	Industrial	Civil Eng.	Approx. UK market £ m
Raw materials					
Cement, pfa and ggbs	280	140	35	245	700
Aggregate	280	140	35	245	700
Reinforcement	30	90	20	110	250
Prestressing					
Admixtures					20
Fibres					
Semi-products					
Ready mixed concrete	220	200	60	320	800
Site mix	175	70	20	85	350
Products					
Precast masonry	340	130	30	200	700
Other precast	290	140	10	260	700
% of market	40	20	5	35	

Figure 2 Markets for concrete industry - 1994

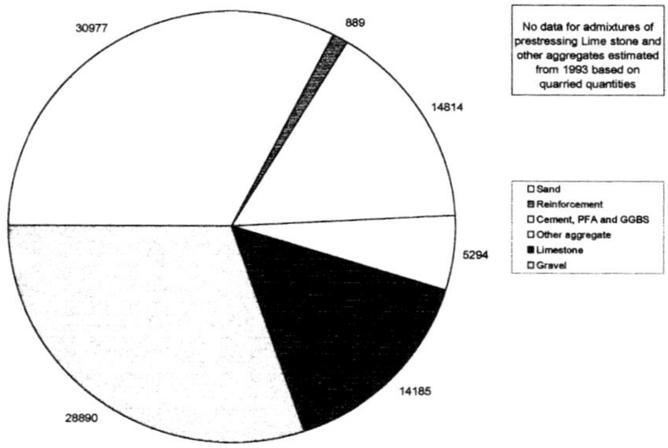

Figure 3 Material used in concrete construction, 1994 (Units in 1000's tonnes)

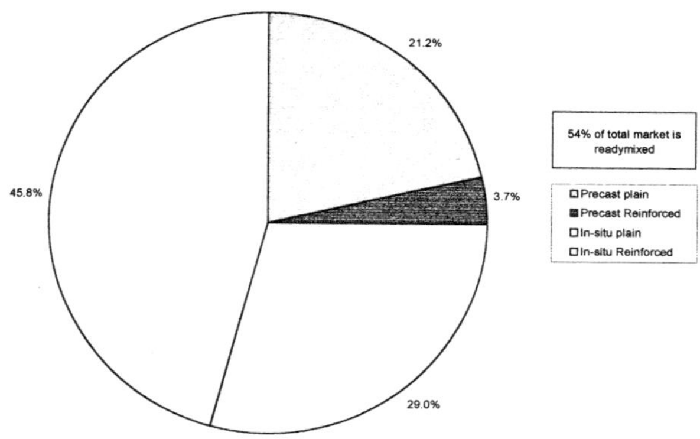

Figure 4 Ratio of types of concrete construction, 1994

SUSTAINABLE CONSTRUCTION

The concept of sustainable construction embraces a number of objectives [2]. Firstly it aims at social progress, recognising the needs of everyone. The objective simply means that everyone should be able to live in a clean and safe environment. The second strand is the effective protection of the environment from global and local threats. This includes the protection of things we value such as wildlife, landscapes and historic monuments. The third aspect is the prudent and efficient use of natural resources. This in turn requires the development of renewables as alternatives to materials and fuels of limited supply. Sustainable development also aims at the maintenance of high and stable levels of economic growth and employment.

These new imperatives have enormous implications for the construction industry but they do not imply that we abandon the way we live now and resort to "cave-man" technology. Rather they demand that we become more efficient and competitive through innovation and better management of our affairs. Everyone in the supply chain of the industry (Figure 1) has a role to play. Considerable research is already underway to facilitate sound decision making. As with all innovations, risks will need to be balanced against potential gains. However, the alternative of doing nothing is not risk-free. It is simply not sustainable. Let us now look at what is possible.

Environmental Considerations

A few bold statistics will show: (a) how environmental considerations are relevant to construction activities, and (b) the scope to contribute to the well-being of the society.

- Energy use in buildings account for between 40% and 50% of the UK emissions of carbon dioxide (CO_2)

- Construction industry consumes around 6 tonnes of material per person per year. This simply dwarfs the amount of material used in any other industry

- About 70 million tonnes construction waste is produced every year in the UK

Thus the scope for good housekeeping, reducing consumption, reducing waste and re-cycling is considerable.

Incorporation of a formal environmental component into construction design will become increasingly important and necessary. Before looking at the steps involved in this process, it will be useful to have an overview of some selected environmental indicators and constructionrelated contributions. This is shown in Table 1 [3].

It will be seen from the table that:

- Energy associated with the provision and use of buildings and structures accounts for about 77% of UK final energy
- Energy associated with construction materials account for about 7.9% of the construction energy (provision and operation)
- Operational energy in running the buildings far outweigh the embodied energy (de) the energy consumption in the production and transport of construction products and material

Life Cycle Analysis

One method of analysing and assessing the environmental impact of a product or service is Life Cycle Analysis (LCA). The International Standards Organisation (ISO) and the Society of Environmental Toxicology (SETAC) are developing guidelines.

Table 1 Seclected environmental indicators and construction-related contributions

INDICATOR	UK TOTAL	CONSTRUC MATERIALS	CONSTRUC PROCESS	BUILDINGS IN USE	DWELLINGS IN USE	TRANSPORT	OTHER SOURCES
Final energy PJ (%)	6283 (100)	386 (6.1)	54 (0.9)	952 (15.2)	1839 (29.7)	1676 (26.7)	1376 (21.9)
Carbon dioxide, Mt (%)	548 (100)	37 (6.8)	4 (0.8)	77 (14.1)	149 (27.1)	145 (26.4)	136 (24.8)
Sulphur dioxide, kt (%)	2718 (100)	151 (5.6)	21 (0.8)	372 (13.7)	719 (26.4)	251 (9.2)	1204 (44.3)
Nitrogen oxides, kt (%)	2218 (100)	231 (10.4)	8 (0.4)	134 (6.0)	259 (11.7)	1356 (61.1)	230 (10.4)
VOCs, kt (1%)	2220 (100)	94 (4.2)	36 (1.6)	27 (1.2)	53 (2.4)	938 (42.3)	1072 (48.3)
Carbon monoxide, kt (%)	4833 (100)	629 (13.0)	3 (0.1)	54 (1.1)	281 (5.8)	3786 (78.3)	80 (1.7)
PM10, kt (%)	250.0 (100)	26.1 (10.4)	4.1 (1.6)	14.5 (5.8)	28.0 (11.2)	71.0 (28.4)	106.4 (42.6)
Heavy metals, kt (%)	4.099 (100)	0.689 (16.8)	0.007 (0.2)	0.120 (2.9)	0.231 (5.6)	0.245 (6.0)	2.808 (68.5)

There are four steps:

1. Definition of goal and scoping: This entails specifying a functional unit (viz) a product or servlce for analysis (e g) supply of one cubic metre of concrete with a specified strength.

2. Inventory analysis: This comprises the quantification of the inputs to and outputs from the system required to supply the functional unit. The inputs will include materials and energy and outputs will include waste and emissions to air and water. See Figure 5 which illustrates the processes involved in the supply of ready-mixed concrete.

3. Impact assessment: Here examining the inventory differences and various environmental indicators on a weighted basis (greenhouse gases, acidification etc) compares alternative functional units.

4. Improvement assessment: Here there are opportunities to improve the cradle-to-grave environment performance.

Some software is commercially available, mainly outside the UK. Currently a multipartner project, sponsored by DETR, is underway. It is in its initial stages. It hopes to define environmental performance in the concrete industry in the UK. It will assess the environmental impact of a variety of concretes and concrete elements. A typical output is shown in Figure 6 [4].

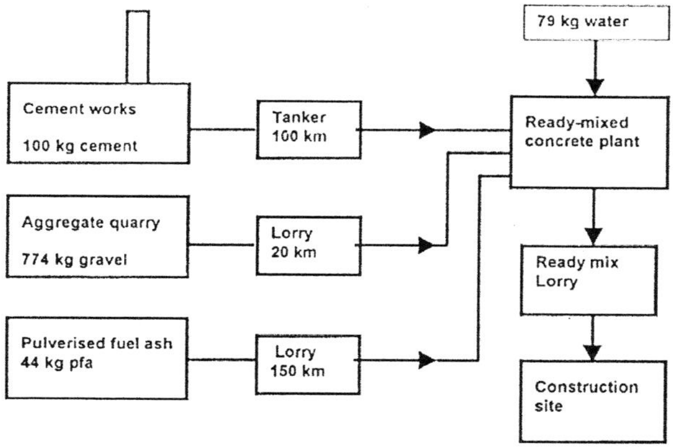

Figure 5 Process diagram for the supply of 1 Tonne of ready-mixed concrete

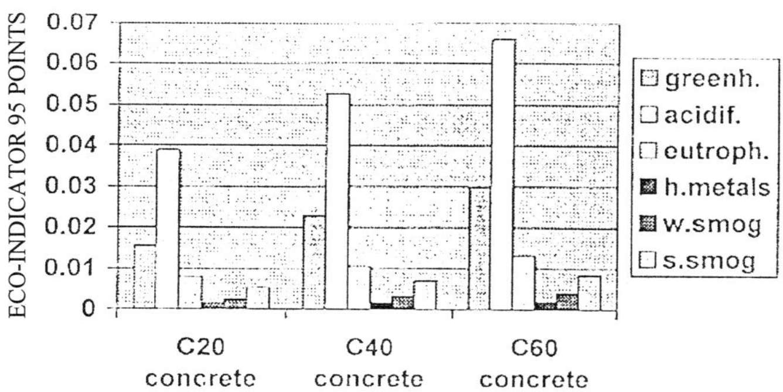

Figure 6 Evaluation of concretes based upon the
ECO indicator 95 method for different indicators

Although Figure 6 suggests that stronger concretes are less eco-friendly, the picture looks very different when durability and service life are taken into account and usage of materials on a "per year" basis is considered. There is a non-linear relationship between service life and the usage of materials. As an example, Parrot has considered the durability and environmental profiles of reinforced concrete columns (Reference 5). Service life was assumed to be limited by carbonation induced corrosion of reinforcement and the formation of visible cracks.

Parrot found that only small increases in concrete quality were required to increase greatly the service life. He evaluated the concrete quality and reinforcement required for different periods of service life for a column of a given height to support a given load. In turn he computed the cost and environmental impacts. The results are shown in Table 2.

Table 2 Material costs and environmental impacts per year of service life

SERVICE LIFE (YEARS)	28 DAY CUBE STRENGTH (N/mm^2)	AMOUNT FOR 1 COLUMN PER YEAR OF SERVICE LIFE		
		COST (PENCE)	ACIDIFCATION IMPACT (kg SO_4)	GREENHOUSE GASES (kg CO_2)
100	39.0 [CEM 1]	55	0.022	2.7
80	38.2 [CEM 1]	69	0.027	3.3
60	37.0 [CEM 1]	94	0.036	4.4
40	35.3 [CEM 1]	146	0.055	6.6
20	32.0 [CEM 1]	299	0.111	13.1
100	47.0 [CEM 2]	48	0.018	2.2
80	45.9 [CEM 2]	61	0.022	2.7
60	44.2 [CEM 2]	84	0.031	3.6
40	41.8 [CEM 2]	128	0.045	5.2
20	37.9 [CEM 2]	272	0.092	10.4

Notes: CEM 1 - Portland Cement
CEM 2 - 30% of the Portland Cement replaced with PFA.

This is a clear demonstration that: (a) higher strengths confer longer service life; (b) cement replacements have beneficial effects on cost and environment; and (c) environmental assessment should be studied in conjunction with service life.

Recycling

One of the main strands of sustainable construction is minimisation of waste, which in part can be achieved through recycling. The waste thus becomes a resource, enabling reduction in demands on natural resources.

Concrete industry already uses recycled material. Few people know that the bulk of reinforcement is produced using scrap metal as the raw material. Recycled concrete aggregates (RCA) are used mainly in road construction and in some applications of plain concrete. More should be possible, starting with situations where performance requirements are modest. To this end DETR has commissioned research in Dundee University. Their initial findings are that: (a) the density of RCA is marginally less than that of natural aggregate (NA); (b) water absorption of RCA was found to be six times higher; (c) RCA concrete specimens were more porous compared with the equivalent concrete using natural aggregate NA; (d) with a mixture of RCA and NA workability was found to reduce with increasing proportions of RCA. Beyond about 30% RCA concrete was found to be "harsh" (less cohesive) leading to higher "bleeding" (loss of water); (e) Compressive strength of concrete with up to 30% coarse and 20% fine RCA was comparable with NA concrete. With higher proportions a reduction in strength of about 10%-15% was observed; (f) Regardless of RCA source, the variability in composite strength of concrete with a range of RCA content was found to be negligible, when compared with corresponding NA concrete.

Some further tests are planned. The specifiers should be receptive to the information and carefully study them. Then they should adopt them for appropriate applications. Initially there will be some inhibitions. Perhaps DETR might consider a demonstration project to improve confidence further. For instance, a structure using recycled materials can be built at Cardington.

Recycling pre-cast building components should certainly be possible. After all, salvaged bricks are re-used. Components will need to be designed and detailed with recycling in mind. Minor modifications of ends of units may be required. However, proper labelling and maintenance of full records will be essential. Other procedures will evolve when the society realises the feasibility of recycling. What is now expensive, will become economically viable as alternatives become relatively more expensive or prohibited by law.

In a similar vein, sites which were considered unviable not so long ago because of contamination, are now being developed more routinely. The government has signalled through their consultative paper "Planning for the communities of the future" its desire to achieve 60% of all new housing on brownfield sites.

Use of cement in this area, to clean and reclaim contaminated sites, has not been fully exploited in the UK, compared to some other countries, such as the USA. The conventional technique of"dig and dump" simply transfers the problem elsewhere. This is environmentally unfriendly and cannot be sustained. There are a number of methods including the following:

- In situ stabilisation and solidification by forming columns of treated soil matrix, using soil mixing augers and injecting cement based grouts.

- The contaminated soil may be excavated, thoroughly mixed with cement and water and the treated soil then placed in layers for compaction in the usual way. Alternatively, the contaminated soil may be excavated and then placed in layers with beds of cement between them and compacted.

- Sludges and sediments in lagoons may be treated directly by mixing with cement slurry which transforms the sediments physically and chemically, resulting in a material suitable for structural fill.

In all these methods, cement based grouts chemically stabilise and physically immobilise the contaminants within a solid matrix. This improves the engineering properties of the ground, minimises leaching and reduces the permeability. Cement based treatment is suitable for a significant number of organic and inorganic containments, particularly heavy metals.

The British Cement Association has taken the initiative to publicise the techniques (6). Designers should consider these alternatives critically and adopt them.

DESIGN

A number of features of the current design practice could be re-examined and more radical approaches could be considered.

Loadings

In commercial developments the client usually stipulates the design superimposed loading to be used. Often these are considerably in excess of the loadings prescribed in BS Loading Codes. The usual reason given is that there may be changes of use in the future which may demand increased loadings. Can we afford this luxury any longer? Will it be more sensible to design the vertical elements and their foundations for some overload but accept that floors may need some strengthening should the need ever arise. Clearly this is a matter for dialogue between the client and the designers. In my experience very few buildings have used the passive overload capacity, in practice. In the context of sustainability such over consumption of materials could be considered wasteful.

The other aspect of design loadings is the application of a uniform load factor (γ_f) for all dead loads, including self-weight of members. In part the load factor caters for variability and uncertainties of loads. Tolerances on the dimensions of members are fairly small and the density of concrete as produced has been fairly consistent. Under the circumstances the case for a reduction of YF is compelling. This is under serious consideration in Eurocodes. In any event, this could be made conditional on verification of construction.

Beeby and Fathilbitaraf [7] have shown that there is scope for reducing the reinforcement in continuous members in bending. This is possible because of (a) redundancy in continuous construction leads to a considerably higher safety level compared to determinate structures and (b) the pattern loadings prescribed in codes are unrealistic. However, if fully exploited this will increase the service stress in reinforcement. Thus serviceability limit state may well control the design. In general this is unlikely to be a problem in beams. Slabs, which account for the bulk of material usage in a building, may in any case benefit from more rigorous serviceability limit state checks for cracking and deflections. This is even more true when realistic service loads are used as in Eurocodes. Currently full characteristic loads (Q_k) are used for serviceability verifications in the UK. Eurcodes use quasi permanent loads (ie) $\psi_2.Q_k$ for reinforced members, where ψ_2 is a factor (less than unity) the value of which depends on the use of the building.

High Strength Concrete

Properties of high strength concrete are now well established through considerable research. On this basis it has been decided to extend the concrete strength class covered in EC2 to C105 (cylinder strength). Many ready mix suppliers are also now familiar with the technology. It is obvious that the higher strength will help reduce the size of compression members. There are many other benefits which are not immediately obvious. High early strength will permit early striking of formwork and speed up in-situ construction. High strength concrete invariably uses superplasticicers and high slump. The increased workability will result in well compacted concrete which in turn will enhance the durability.

In flexural members the increased tensile strength of concrete will tend to reduce deformations. The increased tensile strength could also be exploited in the design of pre-stressed members in some circumstances.

Housing and Concrete

Currently in the UK, low rise housing generally uses cavity walls for the external envelope, timber floors and stud internal walls. Society still generally prefers facing bricks externally. For thermal insulation the inner leaf is generally in lightweight blockwork such as those made out of autoclaved aerated blockwork (AAC). In the Continent the use of concrete in housing is common. Thermal insulation requirements in some of these countries are even more stringent than in the UK. There will be considerable advantage in using reinforced AAC (RAAC) panels for inner leaf and flooring. In this form strength and insulation characteristics are combined. Being pre-cast and lightweight, the structure can be put up with considerable speed. Acoustic insulation use will also be better as the density of floors will be greater than traditional timber construction. It goes without saying that all the components should be properly designed and detailed. This is an area well worth exploring.

Codes

As engineer's work is concerned with public safety, it will always be subject to regulatory control. There are a number of reasons why codes evolved [8]. They include the following: (a) codes enable consistent enforcement of regulatory requirements both geographically and with respect to time; (b) codes can form the basis for tenders and appointments and (c) they provide an agreed basis for design aids and manuals.

As with all standardisation, the question is how far does one take codification? The answer depends upon the quality of the users. If engineers are well educated and highly trained then the code could be a simple set of Principles. In this scenario, every engineer can come up with his own ingenious method for compliance. This is if course pie in the sky! The credentials of the average engineer forces one to conclude that the code should go a bit further than just principles. In fact Eurocodes do just this. They contain Principles with which all designs must comply and also Application Rules, for which alternatives are permitted. Thus Eurocodes are the nearest we are likely to have by way of innovative codes for quite a long time. They are flexible in that they do not inhibit innovation but also provide well proven methods for the routine jobs which are in truth the main stay of the industry.

Specifications

Every designer prides himself with a bespoke specification for every project. When one analyses projects, it is obvious that most processes adopted by the different contractors, irrespective of the origin of the specification, are similar. Standardisation in this area clearly will improve the efficiency of the industry. Like Codes of Practice, it will facilitate clear communication between parties. The recently published "National Concrete Frame Specification" [9], points the way. Part 1 of this document contains all the usual clauses, (eg) materials, workmanship, etc.

It is followed by Part 2, which deals with projectspecific requirements. It lists the information to be supplied to the contractor together with the information to be provided by the contractor. The designer can insert special requirements if any. This seems an excellent start and the industry should discuss where similar approaches may be possible.

QUALITY ISSUES

There is a growing awareness in the industry for the need to achieve good quality construction and getting it right the first time. I will briefly mention three areas which are receiving attention.

Whole Life Costing

Most clients understand that the lowest first cost does not necessarily give the best value for money. Lowest first cost has an appeal when the responsibility for capital and maintenance is not in the same hands (e.g.) different budgets or buildings developed and sold on. Thus a more holistic view is not a consideration. The culture is now changing (slowly) and whole life costs are now considered in some projects. Whereas whole life costing had been used frequently to assess the merits of different design solutions or alternative materials, until recently it had not been fully utilised for assisting investment decisions. It is partly for the reasons cited above and also the methodology had not been fully evolved. Now BCA have prepared a useful report for DETR [10] which links quality to whole life costing. It defines a model for service life at the end of which some intervention is likely to be necessary and provides a method for calculating costs on a normalised basis. When predicting the future there are clearly imponderables but their sensitivities can be tested The industry thus has the tools necessary to take the method further. For bridges, similar methodologies have existed for some time but there the builder and operator are usually the same body. Whole life coatings will be very relevant to PFI schemes.

Certification

There is a desire in some quarters to develop a certification scheme for concrete structures. There is indeed currently a project sponsored by DETR to explore the mechanics of how this might operate. It can only be a good thing. It is possible that only limited operations in the construction process might be covered by this scheme. It will still be better than nothing. After all we have similar schemes for things such as steel reinforcement, ready-mixed concrete etc. When designers and clients demand these, the industry will deliver them.

Operative Training

It is absurd that at this juncture in the history of concrete we still do not have a recognised and proud tradesman called "The Concretor". This is largely the fault of the designers and clients for not insisting on the work being carried out by qualified tradesman.

However, the trend is now changing. Major clients are now demanding this and the contractors are beginning to respond. There are also moves afoot to support and enhance some of the NVQ's to facilitate the culture changes needed to achieve quality.

CONCLUDING REMARKS

The survival of the industry demands innovation and radical thinking. I have touched upon a number of current issues exercising the concrete industry. Sustainable construction will force us to consider the environmental impact, use of recycled materials, and minimisation of waste. Design and quality issues will also contribute to sustainability. In design use of more realistic loads and load factors should lead to economies and contribute to sustainability. Potential use of concrete in housing needs serious consideration. Specifications could be standardised to achieve clear communication. Quality of concrete construction will be improved when the significance of whole life coatings is fully appreciated. Certification schemes for contractors and the use of qualified operatives will further underpin quality.

REFERENCES

1. DEPARTMENT OF THE ENVIRONMENT Concrete 2005: DOE's Innovation and Research Strategy for the Concrete Industry Sector - June 1996
2. DEPARTMENT OF THE ENVIRONMENT, TRANSPORT AND THE REGIONS. Opportunities for change. Consultation Paper on a UK Strategy for Sustainable Construction – DETR
3. PARROTT L, BCA: Environmental Perspective on UK Construction Materials: (CIB World Building Congress 1998, Gavle, Sweden)
4. PARROTT, L. BCA: Environmental Performance of Concrete - A Partners in Technology Project
5. PARROTT, L. BCA: Durability and environmental profiles of reinforced concrete columns, 6th LCA Case Studies Symposium SETAC-Europe 1998
6. BRITISH CEMENT ASSOCIATION. Recycling Contaminated Land using Cement
7. BEEBY, A W, FATHILBITARA, F "The Design of framed structures - A proposal for Change" Int Conference. "Concrete in the Service of Mankind" University of Dundee, June 1996 pp 495-506.
8. NARAYANAN, R S: "Structural Codes of Practice - A UK point of view" - Structural Engineers World Congress, July 1998 p 318-2.
9. National concrete frame specification for building construction - Construction Research Communication, 1998.
10. BRITISH CEMENT ASSOCIATION. Relating quality to whole life costing - Report to DETR - July 1998.

THEME ONE:

NEW PRINCIPLES

Keynote Paper

RADICAL CHANGE OR INCREMENTAL IMPROVEMENT

A W Beeby

University of Leeds

United Kingdom

ABSTRACT. The question of innovation in construction, and particularly in concrete construction is addressed. The reasons for innovation are considered and case studies are briefly described to illustrate different ways in which innovation has been attempted in construction and the results of failure and success. It is pointed out that there is a high degree of risk inherent in innovation. A simple conceptual model relating the risks to the degree of innovation is introduced. This illustrates how incremental innovation may, in the long term, achieve changes as large as single major innovative jumps with much lower risks. A broader look at innovation suggests that process innovation may be inherently less risky than product innovation and that incremental innovation may have considerable motivational advantages.

Keywords: Innovation, Incremental innovation, Risks, Examples.

Professor Andrew W Beeby is Professor of Structural Design at the University of Leeds, UK. He has specialised in the development of design methods for reinforced concrete and is a member of both UK and European code drafting committees. Over the past 35 years he has carried out research into many aspects of the structural behaviour of concrete members; a particular area of interest has been the prediction of crack widths and deflections. Since 1990 he has chaired Permanent Commission 2 - Member and Material Modelling of the Comite Euro-International du Beton. He is currently a member of council of the Institution of Structural Engineers

INTRODUCTION

A keynote paper, in my view, should be a paper which tries to cover a broad field, possibly in a thought provoking way. It should not be an individual research paper on a particular topic. This paper therefore aims to explore some general issues related to innovation in construction, particularly with reference to concrete construction. The issues are not explored in depth so the paper should not be viewed as a 'state of art' paper either. Issues of innovation have been considered in depth by researchers, but more particularly in the construction management field than the field of concrete construction (see, for example, papers in (1)).

Usually, the discussion of innovation centres of the perceived low rate of innovation in Civil Engineering and the need to innovate more. The question, put another way is: "why is the Civil Engineering industry so resistant to change and innovation?" The view of many practising engineers is believed to be roughly that "if it works, don't fix it". Research, which could help to inform innovation, is often viewed with great suspicion and is thought of as merely making design and construction more complicated.

Since I don't want my arguments to be construed as a suggestion that we shouldn't innovate, let me make some initial general points.

Innovation has taken place continuously, though at varying rates, throughout the whole period of human existence. Indeed, the ability to innovate may be what sets Homo Sapiens apart from other animals. Innovation will continue and future ages will be as different from the present as the present is from the past. It is interesting to speculate what views the inhabitants of past worlds would have actually had about innovation, had they been asked. My guess is that most people asked would have said that life seemed pretty reasonable on the whole though, if someone could think of a way, there were one or two things that could be improved a bit. In all ages, however, there were some people who had visions of major change. Some, particularly in the period since the Industrial Revolution, will actually have helped major changes occur; others will probably have been burned at the stake or died paupers. Innovation is not easy, it is not popular and it is risky.

The argument that innovation is natural to mankind is not an argument that can be used to explain why industry has to innovate and it may be worth clearing the ground by making clear why innovation is necessary. This may appear obvious to many but it rarely seems to be stated.

In 1776 Adam Smith published his groundbreaking book "The Wealth of Nations" which set out the basic principles of the capitalist market economy. His analysis demonstrates how, in a society where there are a large number of manufacturers producing a particular product, competition will ensure that the appropriate quantity of the product was produced at a price which was close to the minimum at which it was possible to operate.

Industry must constantly search for means by which a better return can be achieved than this bare subsistence level and there are various possible approaches to this. For example:

1. forming a cartel to fix prices at an artificially high level.
2. buying out the competition and hence forming a monopoly which can fix its own prices
3. more aggressive advertising and marketing to make ones product appear more attractive than that of the competition
4. increase the size of the market by some means
5. find a means of making the product more cheaply without reducing quality
6. find a way of producing a product which meets its purpose better and hence permits the charge of a higher price.

1 and 2 above act to subvert the market and governments have tried, with varying success to make these illegal, or at least control the possibilities. 3 is used extensively in the retail market but is of limited value in the construction industry. 4 has been used to great effect and the British Empire, which provided a cheap source of raw materials and an almost bottomless market for cheap manufactured goods, was the basis of our wealth for about a century until the early years of this century. Construction, of course, still works very extensively abroad but the market is open to competition from many other countries and hence no longer allows the charge of rates much above the subsistence level. This leaves 5 and 6 as the most reliable approaches to gaining an attractive return on money and effort; these two cover Innovation. Competition produces a treadmill which we cannot afford to get off. If we innovate and others do not, then we gain market share and the others lose. If others innovate and we do not, then we lose. In the absence of a moratorium on innovation, we can only maintain a reasonable return on capital employed by innovating.

A variation to this answer can also be drawn from natural selection. When the environment within which a species operates changes, then either the species changes to be able to operate successfully in the new environment or the species becomes extinct. The parallel here is that, where the environment within which construction is carried out changes, survival may well require innovation and this innovation may be major.

Innovation here and throughout this paper is taken to be any change in the technology of an industry or in the way the industry operates with the objective either of improving the product or of reducing the manufacturing cost of the product.

To pursue the discussion further, two case studies will be briefly presented.

CASE STUDIES

Two case studies of innovation will now be presented which, it is hoped, will illustrate some further points about innovation. The first of these is historical and might not now be considered to be 'civil engineering', though at the time it was. It also has nothing to do with concrete. The second does deal with concrete but may well not have been noticed as "innovation" by concrete practitioners.

Case Study 1: The Attempt by Brunel to use the Atmospheric Railway.

The following is largely drawn from L. C. Rolt's biography of Brunel (1).

Brunel was an innovator par excellence in an age when the development of the railways and sea transport was requiring engineers to solve new problems which could not necessarily be solved by modifying earlier solutions. Many of Brunel's projects were highly innovative for the time and were successful but some were not; the atmospheric railway was one that failed. It is the reasons for the failure which are really of interest here.

As Engineer to the South Devon railway, Brunel had the job of building a railway from Exeter to Plymouth. The great Western's broad gauge empire had already reached Exeter and the West Country was the next area to develop. There were, however, a number of major engineering problems to be solved within tight financial constraints. These constraints mainly arose from the rather limited trade for the railway which could be expected due to the relatively undeveloped and lightly populated country west of Exeter.

The engineering problems related to the terrain which the railway had to be build across, which was much more hilly than other routes that Brunel had been involved with. Steam locomotives rely on friction between the rails and the wheels to enable them to pull load. To obtain this friction, the engines have to be heavy. Furthermore, the steeper the gradient, the greater force the locomotive must develop and the heavier must be the engine. All this weight required even more power to be developed. As a result, railways are not good on gradients and considerable efforts are made to limit them. Brunel had been particularly assiduous in limiting the gradients on his earlier railways, such as the Great Western line from London to Bristol and this enabled very fast and economical running by the standards of the time. He had thus been prepared to invest heavily in the capital cost of his lines in order to reduce running costs. This capital was not available to him for the South Devon Railway and the proposed route for the line had a number of very severe gradients which Brunel doubted the ability of the locomotives of the day to handle economically. He therefore looked for some other form of motive power and hit upon the concept of the atmospheric system. This was not a new idea and a number of short lengths of atmospheric railway had been installed. A short length of about $1^1/_4$miles had been installed in West London and had produced very encouraging results. A somewhat longer length of line but up a gradient of 1 in 128 was constructed in Ireland. On one occasion a single carriage ascended the line at an average speed of 84 miles per hour; possibly a world speed record at the time. Brunel visited both these railways and his assistants carried out trials on the Irish line to establish the economics of the system. The system worked by laying a pipe with a slot along the top between the rails. This pipe was connected to a steam pumping engine which extracted air from the pipe. The train had a piston fixed below it which fitted inside the tube and was connected to the train by a bar which passed through the slot. The slot was sealed with a flap valve which was automatically lifted by the train as it passed. The train was thus sucked along the line (see Figure 1). The great potential advantage of the system was that the power source was static and did not have to be carried by the train. The power required was thus reduced and the problem of friction became irrelevant, nor did fuel have to be carried. The Irish experience had shown that the system had great potential for handling gradients. There were some mechanical problems to be overcome but Brunel felt that these were unlikely to be insuperable. Brunel therefore proposed the Atmospheric system for the South Devon line and this was accepted by the directors. The line was constructed and the atmospheric system installed during 1846 and 1847 and it was in service by September 1847 over parts of the line. Construction had not been without its problems and considerable time and extra expense had been expended on the caulking of the pipe and on the engines.

By 1848, however, it was clear that the system was not operating well and constant breakdowns were occurring. There were various reasons for these but the overriding problem was with the flap valve that sealed the pipe. Leakage of this valve meant that the fuel consumption necessary to keep the required level of vacuum in the pipe was excessive. Furthermore, water soaked into the leather from which the hinge of the valve was made which, when it froze, caused the hinge to break off. By June 1848, it was concluded that the maintenance of the flap valve was impractical. This, together with the unexpectedly high running costs led to the abandonment of the system and the conversion of the whole line to steam locomotive operation.

Where did this attempt at innovation go wrong? Brunel had a clear technical and economic problem to solve and he appeared to have found a rational solution. He did not accept the idea blindly, he studied the available experimental lines that had been set up and, indeed, carried out tests to establish the technical and economic parameters. What more should he have done? With hindsight we know the answer to this but were his decisions unreasonable at the time he made them with the information he had available? He was aware that the experimental lines had had difficulties with the flap valve but concluded that this was, mechanically, a fairly simple device and that the difficulties could be overcome by a little development work during the design and construction. It must be true of most developments that the solutions to problems are continually being developed during manufacture and have not all been fully solved prior to the decision to go ahead. Brunel's problem was that he had, unknowingly, jumped ahead of the current knowledge of materials science; the failure was basically a materials failure. At that time, the means to produce a suitable valve simply didn't exist and this resulted in the failure of the system. Innovation requires a leap into the unknown and it is always possible to crash on landing. In this case, Brunel crashed.

A consequence of this spectacular failure was that the atmospheric system was viewed as an aberration and an example of technological idiocy from that time on. Hardly anyone has considered its possibilities since or considered whether alternative designs which got round Brunel's problems could have been devised. Brunel's failure killed the concept but was it really so mad? Nowadays, the problems which faced Brunel can be solved and, indeed, I have actually travelled for a mile or two on an atmospheric train. The valve had a different configuration and was made of Neoprene. Leakage was less of a problem because the pipe was of a much greater cross-sectional area and hence the air pressure could be much lower. Brunel could not obtain larger diameter pipe that the 25" he used.

Case Study 2: Changes in Reinforced Concrete

Figure 1 shows how the design stresses in service in reinforcement have changed over the years. The solid line shows, for codes where permissible stresses were used, the permissible stress and, for CP110 and BS8110, an estimate of the stress likely under service loads calculated on the basis of a modular ratio of 15. The dotted line indicates the service stress that might be expected at a critical section where the maximum permitted amount of moment redistribution had been carried out.

It will be seen that, over the last 50 years or so, the stress level with no redistribution has increased by about 160% and the stress level with redistribution by 210%.

Alternatively, from an economic point of view, the required quantities of flexural reinforcement have been reduced by more than 60%. This change has been gradual, with an upward movement at every revision of the design codes. This would probably not even be recognised as innovation, however, the overall change to the amount of reinforcement used and the likely performance of reinforced concrete structures is very large indeed; larger than that achieved by most recognised 'innovations'. A slightly closer look at the history of these changes is instructive. For a beam of given span and overall section dimensions supporting a given loading, one might roughly expect the deflection to be proportional to the solid line while maximum crack widths could be expected to be proportional to the dotted line. The large increase in stress in 1948 arises from the permission to use high yield steel at a permissible stress of 27,000 lbs/in^2 while earlier documents effectively applied mild steel stresses for everything. The consequences of this increase in stress on deflections was not recognised in the 1948 version of CP114, which contained no limitations on span/depth ratios. This was introduced in 1957, presumably after experience with the higher stresses introduced in 1948. The code still, however, required no check on cracking. This was introduced in 1972, together with a considerably more rigorous treatment of deflections following further significant rises in steel stresses. It will be seen that small changes have been made, experience has suggested consequential measures which were needed to deal with these changes which were included in the next code revision, and so on.

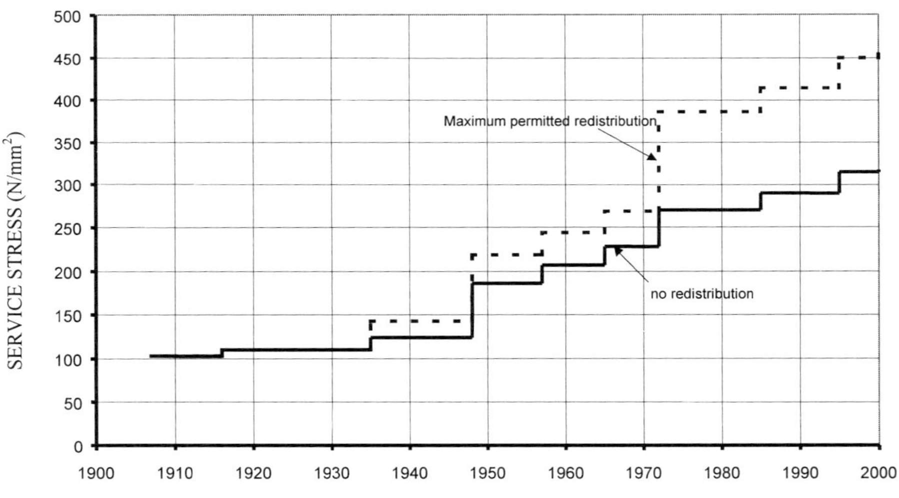

Figure 1 Variation of service stress in reinforcement over time

In the case of steel stress levels in reinforced concrete, large changes have been introduced relatively painlessly. It is interesting to imagine what might have happened if this change had been introduced in one leap in 1948. Up to this time designers had never had to consider the deflection of their structures or cracking.

Would they have realised that the increase in stresses changed the situation quite drastically or would we have had a series of expensive disasters in the 1960s which could have set back concrete construction by many years? The success of this incremental innovation may be compared with the results of other more sudden innovations which were introduced in the 1950 and 1960s such as industrialised multi-storey buildings.

A third case, which will be covered only very briefly, is included to show that the construction industry is not incapable of major innovation when necessary. During the late 1950s and early 1960s a huge reconstruction exercise took place in the UK. This was to rebuild city centres destroyed in the war and to catch up on some 30 years of redevelopment which had not taken place due, firstly to the depression of the 1930s and secondly due to the war and the ensuing lack of resource. An immense building programme was carried through and much innovation had to take place to make this possible within the resources available. This was the period of high rise flats, industrialised building, designed concrete mixes, the first major usage of high strength reinforcement, the introduction of ultimate strength design rather than elastic methods, and much more. In other construction areas, this was the period of major motorway building, which resulted, inter alia in the development of the use of standard prestressed bridge beams which became the dominant form of bridge construction for a quarter of a century. As someone who entered the industry at that time, I can say that it was a time of great enthusiasm for new methods and materials and an exciting feeling of moving into a new scientific era.

Unfortunately, quite a number of the innovations introduced in this relatively short period were found to have technical problems not forseen at the time. The collapse of Ronan Point was probably the first major factor which crystallised the feeling that all was not well. This was followed by a succession of other problems; collapses at Camden, Stepney and Aldershot, corrosion due to the use of Calcium Chloride, the conversion of High Alumina Cement, Alkali-Silica Reaction and corrosion due to de-icing salts are the major ones. Possibly these failures, and the damage they did to the self confidence and the public image of the industry are a significant cause of the caution which clients and the industry now show towards innovation. Concrete was not the only material to suffer but it suffered particularly badly and has by no means fully recovered. Quite a proportion of the odium with which concrete came to be viewed was not directly the fault of the concrete industry but of planners as many, possibly most of the problems with multistorey blocks of flats were social rather than technical in origin.

Did the concrete industry make mistakes in this period which it could reasonably have avoided?

With hindsight, almost all the technical problems could have been quite easily avoided at little cost. Unfortunately hindsight is not a good guide to the environment in which the innovations were developed and the state of understanding of design and of materials of those introducing the innovations. A great deal of effort went in to establishing the causes of the problems and the understanding of these causes is now part of our culture; it is impossible to imagine what it was like working before this understanding was general. The pressure on the industry to deliver on the promises of the politicians should also not be underestimated; nor the real needs of the country at that time and the supply and resource problems under which the industry was working. Probably the pressure of circumstances made a considerable degree of failure inevitable.

These case studies are intended to highlight several points:

(a) Innovation is accompanied by risk. Our education and training as engineers does not appear to enable us to identify all the problems arising from our innovations where these lie outside the current technology.

(b) Large innovations are accompanied by large risks while successive small innovations are accompanied by smaller risks which can be handled without major consequence.

(c) A consequence of failure is often an over-reaction against the particular technology which may set back its development for decades.

(d) It is incorrect to say that the concrete industry by virtue of its organisation cannot innovate. In the past it has but has got its fingers burned.

A MODEL OF INNOVATION AND RISK

If we can propose an analogy with some more readily understood problem, it should be possible to establish at least the form of mathematical relationships to define the risk associated with innovation. A model which seems appropriate is the following.

Assume that you are standing on highish ground and looking across an extensive plain towards distant high ground called ‘success’. You can roughly discern a road network between your present position and success and your task is to propose a route. We are in an earlier age when travel by road was beset by dangers, for example, robbers, washed out bridges, marshes, plague etc. Some of these may be seen from your vantage point and can be avoided but many cannot, being hidden by low hills, woods etc. The journey is thus beset by risk; can we say anything mathematical about this risk? If it is assumed that there is a constant risk per unit distance travelled that you will meet an insuperable obstacle, then the probability of not arriving may be written as:

$$P_f = 1 - k^x$$

where P_f = the probability of arriving
k = a constant defining the level of risk per unit distance
x = distance

This relationship is shown as a broken line in Figure 2. Consideration of our model can, however, lead to some modification of this relationship. When considering a route from our vantage point, the road from where we are standing is clear and we can see, possibly for a mile or two, that there is no danger. With increasing distance, the less we can see and the lower is our ability to discern possible dangers. The risk per unit distance travelled is therefore not constant but increases with distance. This will result in a relationship between risk of not arriving and distance which is initially flatter than the relationship above but becomes steeper than the curve above with increasing distance. This is shown as the solid line in Figure 2.

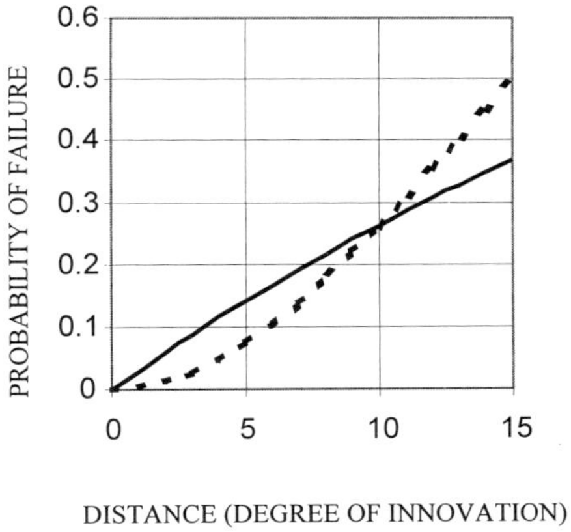

Figure 2 Conceptual relationship between risk and degree of innovation

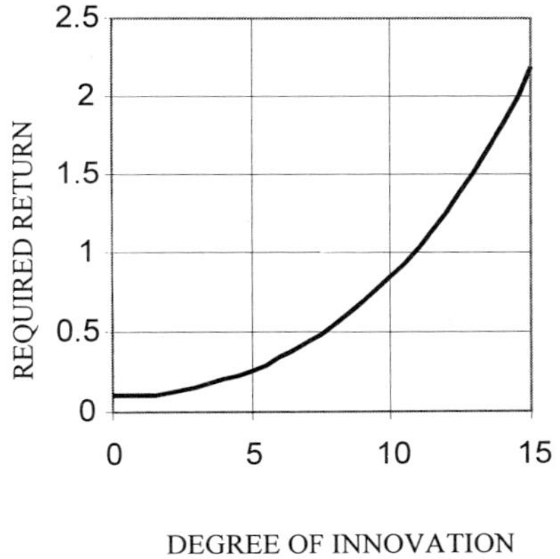

Figure 3 Conceptual relationship between degree of innovation and required return

If the model proposed above is a reasonable analogy for innovation then the solid line in Figure 2 can be taken to indicate the form of relationship that can be expected between size of innovative leap (degree of innovation) and risk. Actual figures cannot be put to this relationship unless a means exists to measure the 'degree of innovation' inherent in a particular scheme and no method currently comes to mind for achieving this. Nevertheless, Figure 2 provides enough information to enable the argument to be carried forward a stage. Innovation would not be considered unless it was likely to yield economies over, or greater gains than current methods. The prospective innovator will have to be able to convince his clients that this gain is worth the effort and that the probability of achieving the prospective gains are high. As in all gambles, greater risks will be countenanced if the potential prize is great. It is quite easy to put some simple statistics to this concept, leading to the relation:

$$g > (pL + g_o)/(1-p)$$

where:

g	=	the gain required if the innovation is to be financially worthwhile
p	=	the probability of not succeeding
L	=	the loss incurred in the event of failure
g_o	=	the gain which would be expected without innovation

This relationship, together with Figure 2 may be used to give an indication of the gain which it will be necessary to be able to forsee in order to justify a given level of innovation. This is done in Figure 3. It will be seen that the return required to justify an innovation rises exponentially with increase in the degree of innovation.

Clearly, the risk varies with the nature of the industry or technology considered and the means that may be adopted to reduce the risks. A few examples will illustrate this.

In many industries it is possible to make prototypes of new designs and test these thoroughly prior to initiating full-scale production. This will be done in combination with extensive market research. In other cases such as the manufacture of drugs, large scale trials of proposed drugs have to be carried out before it is released onto the market. Both these procedures help greatly to reduce the unknown factors which may influence success or failure, but will never eliminate them completely. Research can also help to avoid technical problems. All these options will cost money to carry out but will reduce risk. The nature of the industry will define what approaches are appropriate. The problem with construction is that projects are one-off and the use of prototypes is not generally feasible. Furthermore, while research can solve technical problems, it is not necessarily easy to see how to commission research to identify unknown problems. The construction industry is, in any case, not well structured to commission or to use research. Construction projects are also often very large and the costs incurred in the event of failure tend to be very large indeed. All these factors, and the generally rather low returns on construction work, act to discourage major innovation. The best option in such situations would appear to be incremental innovation and, as has been seen, this happens. It is possible that a useful parallel with the world of nature can be drawn here. Natural Selection results in species developing near optimum strategies for exploiting the available resources. This does not require any guiding intelligence; it is merely the result of the successful approach conferring a higher probability of survival. Is it possible that incremental innovation with generally small steps happens in Construction simply because this is the most successful strategy in the environment within which construction is carried out?

Even within construction, or within the concrete industry there are areas where larger scale innovation may be possible. Writers on the subject of innovation (eg 2) divide innovation into two categories: Product innovation and Process innovation. The second of these categories can be further subdivided into Hard and Soft Innovation. The differences between these categories can best be illustrated by examples. Product innovation could include development of a new, innovative material or form of structure. Hard process innovation would include developments in the technology of production, for example the development and use of new plant or equipment. Soft process innovation covers such things as procurement methods, management or organisational changes. The risks associated with each of these can be considered. The case studies outlined above are cases of mainly hard product innovation and, as discussed, are risky. Soft process innovation could be considered to be associated with very much lower risks; for example, if an alternative management system is introduced and it doesn't work very well, then the losses incurred are likely to be very limited while the gains if it works can be very considerable. Hard product innovation can be expected to be intermediate between hard product and soft process innovation. From the case studies, it has been seen that it is useful to differentiate between two types of innovation: the major leap forward and the incremental change. Lahdenpera, Koskela and Tanhuanpaa (3) refer to these as 'innovation' and 'incremental improvement' respectively. It has been argued that the risks resulting from incremental improvement are substantially lower than those from innovation. This categorisation permits the production of the conceptual matrix shown in Table 1 below.

In looking at this table, it should be pointed out that it only shows the negative side; the possible losses if the innovation fails and the risk of this happening. The possible benefits are not considered. Overall, these may be large in all categories but it does not appear to be possible to say that one category is intrinsically likely to produce more benefits than another; it depends on the particular innovation. Looking at this table, it may be no coincidence that the most active area for innovations at present is the soft process innovation field.

Table 1 Risks related to various types of innovation

		INNOVATION	CONTINUOUS IMPROVEMENT
Product innovation			medium/low risk of failure potential losses limited
Process innovation	Hard	highish risk of failure lowish potential losses	low risk of failure low potential losses
	Soft	highish risk of failure low potential losses	very low risk of failure minimal potential losses

Lahdenpera, Koskela and Tanhuanpaa (3) discuss the soft process innovation area in relation to the relative merits of innovation and continuous improvement and Table 2 is taken from their paper. While this table is related specifically to process improvement, much of what it says must be applicable to other categories of innovation. In particular, the involvement of everyone in the innovative effort and the use of the know-how of the acts very positively to improve the commitment of staff to changes staff whereas the imposition of radical solutions from outside specialists can build up positive resistance to change. Lahdenpera, Koskela and Tanhuanpaa (3) make the point that an organisation committed to continuous improvement will be much better fitted to handle major innovation when this becomes necessary.

CONCLUDING DISCUSSION

The paper has aimed to show that innovation is essential to survival in the modern working environment but that there is real risk attached to innovation. It is suggested that this risk may be reduced by innovating in relatively small steps with a consolidation stage after each step. Though slower than trying to innovate in greater leaps, major change can still be effected by this route. It is also suggested that innovations in the process of design and construction may be much less risky than changes in the technology so that greater changes may be attempted.

The pay-off from such changes are likely to be as great as those from changes in the technology. A reasonable route forward would therefore seem to be to look for incremental changes in the technology while attempting more radical changes in the organisation of the processes of design and construction. It can be argued that this is exactly what is happening.

This ideal approach may, however, be overtaken by external factors which may force the pace of innovation. Experience shows that, in such circumstances, we have to be very careful indeed if we are not to court disaster.

Disaster arises from our apparent inability to identify the full consequences of change in advance. The question which should be explored is: are there ways by which our ability to foresee the consequences of change can be improved? The following possibilities may be suggested.

Table 2 Comparison of innovation and continuous improvement
From Lahdenpera, Koskela and Tanhuanpaa (3)

	INNOVATION	CONTINUOUS IMPROVEMENT
Focus	Efficiency of conversions	Efficiency of work flow process
Goal	Leaps in efficiency	Small steps, details, fine tuning
Involvement	Company and outside specialists, Champions	Everybody in company
Time frame	Intermittent and non-incremental	Continuous and incremental
Technology relied upon	Outside technological breakthroughs, new inventions, new theories	Internal know-how, best practice
Incentive	New superior technology or need for capacity extension	Overcomes constraints in variability reduction or cycle time compression
Practical requirements	Requires large investment, but little effort to maintain.	requires little investment, but great effort to maintain
Mode of action	Scrap and rebuild	maintenance and improvement.
Transferability	Transferable: embodied in individual equipment and related to operating skill.	Primarily idiosyncratic: embodied in system of equipment, skills, procedures and organisation.
Effort orientation	Technology	People

Increase research

Clearly, more research will increase the pool of knowledge available to innovators simply to increase research is not enough. The pressures from industry and funding bodies are for research directly related to current methods and this is part of the reason why our vision of the consequences of the new is so limited.

For example, prior to about 1976 durability of concrete structures was not seen as a problem and there was minimal interest in research in this field. Once it was identified as a major problem, it probably took approaching 10 years before the research expertise had built up sufficiently to provide useful answers. Blue sky research, which could increase the pool of knowledge more broadly is likely to provide a much more fruitful pool for assessing the consequences of new ideas but is not strongly encouraged in the current environment. A further major problem is the dissemination of the information from research so that those who are involved in innovation are aware of the results. In this area there are faults on both the research side and the industry side.

The rules governing the obtaining of research funds actively encourages the publication of research in a form, and in journals which are not easily accessible to those in practice and actively discourages activity aimed at making the results more accessible.

In Design and Construction there is a strongly anti-academic culture which means that the common run of practitioners deliberately avoid reading anything that might be considered as 'academic'. There is a need for a major culture change both within the research community and within industry if the full value of research is to be realised.

Better Education

Too much of current education is concerned with teaching how to operate the system (ie. how to apply design rules) rather than imparting understanding of the technology. Again, the fault is not all on one side.

The professions, through the institutions and the Engineering Council have been pressing for broader degrees, which inevitably means shallower coverage of all subjects. Students find it much easier to learn the 'how' rather than the 'why' and their objective is to pass their exams with minimum effort and furthermore, academia finds the 'how' much easier to teach than the 'why'. Again, a revolution in attitude from all concerned is required.

Demonstration Projects

Like prototypes, these are an excellent way of establishing at least some of the problems before trying out an innovation on a major project. It has been used with success, for example, by Maunsells in the development of the use of fibre reinforced composites (4). The European Concrete Building Project at Cardington is another example of an ambitious demonstration project.

These are just three possibilities and there are clearly others which could help such as Risk Analysis. Clearly, however, if we are to avoid some of the problems which have occurred in the past with innovation in concrete design and construction, major thought is going to have to be applied to the culture of the industry.

REFERENCES

1. ROLT, L T C, Isambard Kingdom Brunel. Longmans, Green and Co Ltd 1957

2. DAVIDSON, C H AND ABDEL MEGUID, T A, (Ed).Procurement - A key to innovation. Proceedings of the CIB W92 Symposium on Procurement, May 1997. CIB Publication 203.

3. LAHDENPERA, P, KOSKELA L. AND TANHUANPAA, V-P, Should we strive for innovation or just run our projects well? Paper in Davidson, C. H. and Abdel Meguid, T. A.,(Ed).Procurement - A key to innovation. Proceedings of the CIB W92 Symposium on Procurement, May 1997. CIB Publication 203.

4. HEAD, P R, Effective research and development for future innovations in the construction industry. Structural Engineering Worldwide 1998. Elsevier Science Ltd.

CONCRETE ON THE BASIS OF INDUSTRIAL WASTES FOR ECOLOGICAL PROTECTION

P V Krivenko **J V Skurchinskaya**

G V Lisenko **P A Korchagin**

Scientific and Research Institute on Binders and Materials

A G Gelevera

Kiev State Technical University of Construction and Architecture

Ukraine

ABSTRACT. Based on the concept that radioactive waste can be chemically fixed by using slag-alkaline binders to from a mineral-like artificial stone, this paper demonstrates different techniques for immobilization of various levels of radioactive waste. Compounds are characterized by a high content of combined radioactive metal (up to 29% by mass), water stability, a low leach rate (less than 10^{-5} g/cm^2 per day) and profitability.

Keywords: Chemical fixing, Geocement, Immobilization, Leach rate, Mineral-like stone, Radioactive waste.

Professor Pavel V Krivenko is the director and scientific supervisor of the Scientific and Research Institute on Binders and Materials (SRIBM).

Dr Janna V Skurchinskaya is the chief of department of binders, SRIBM.

Dr Alexander G Gelevera is a Lecturer in Concrete Technology, Kiev State Technical University of Construction and Architecture, Ukraine, Kiev

Ms Galina V Lisenko is a post-graduate research student at the SRIBM.

Mr Pavel A Korchagin is a post-graduate research student at the SRIBM.

INTRODUCTION

Today, as a result of activity of man on the planet, a huge amount of various wastes have been accumulated, which represents a serious danger both for human beings and for nature. Radioactive industrial wastes of various degrees of activity are among the most dangerous substances from the ecological point of view.

Amount of high active wastes of a nuclear power plants is less than 1%. The low active and medium active wastes form the greater part of wastes, which contain radionuclides with a half-life of about 30 years and which to be stored during 300 years.

The existing practice of the liquid radwastes reversion at nuclear power plants is to store them in storehouses in a liquid condition. This method of storage can be considered only as a temporary measure.

The developed new approach to solving this problem is based on the creation of a long- lived mineral-like matrix owing to a chemical fixation of radionuclides within it.

By the investigations held it was established that with regard radioactive wastes with low and intermediate level specific activity, the priority should be given to a newly developed class of cementitious materials, namely: *geocements*, which are themselves hydraulic binders based on alkali metal compounds, and a mineral stone made with them is found to feature a set of high properties such as high strength, water stability, etc. The *geocements* have been developed and introduced by the V.D.Glukhovsky Scientific Research Institute on Binders and Materials, which has been involved in the synthesis of alkaline cementitious systems since 1957 [1-5].

Synthesis of zeolite-like hydrate neo-formations (hydration products) in the geocement stone, which are known to possess excellent sorption ability, is an additional factor to provide physical fixation of radioactive elements. Hazardous elements, thus, not only form an artificial stone structure, but turn out to be locked in a threefold lattice of the zeolite matrix, which has great amount of energy unsaturated cavities[8].

The *geocements* are the products of chemical interaction of alkali metal solutions with clayish minerals (kaolin, montmorillonite etc.) or with industrial wastes, by-products (granulated metallurgical slags, ashes, etc). In the first case, the hydration products are found to be analogs to naturally occurring zeolites of $Me_2O{\cdot}Al_2O_3{\cdot}(2\text{-}4)SiO_2{\cdot}2H_2O$ (hydronepheline, natrolite, kalio-philite, analcime, etc.), where M - an alkaline element [3,6]. When the use is made of metallurgical slags, a mineralogical composition of the hydration products (*slag alkaline cements)* comprises also tobermorite-like low-basic calcium hydrosilicates, hydrogarnets, alkaline- and alkali-earth hydroaluminosilicates.

The mentioned mineral-like artificial neo-formations feature high durability, water resistance and strength properties. It was found that some similar compounds contained in the structures of ancient concretes, thus explaining a great durability of these concretes, which remained undamaged until now. During the immobilization of radionuclides of alkaline and alkaline-earth metals, the latter enters into a structure of alkaline-earth stone as active component and are reliably chemically combined in it [8-10].

EXPERIMENTAL DETAILS

Materials

For immobilization purposes the real liquid radioactive wastes from nuclear power plants with reactors such as WWER and RBMK were used, which belong to wastes with medium activity, whose compositions are represented in Table 1.

Table 1 Compositions of liquid radioactive industrial wastes in nuclear power plants

INDICES	UNITS	NUCLEAR PLANT, REACTOR TYPE	
		Khmelnitskiy WWER	Chernobyl RBMK
Density of a concentrate of cubic residual	Kg/l	1.32	1.23
Content of salts in a concentrate	G/l	453	365
Hydrogen index	pH	11.9	13.2
Mass concentration of the elements:			
Na^+	G/l	196.1	119.2
NO_3^-		1.7	206.1
BO_3^-		140.4	-
Cl^-		2.0	3.5
SO_4^{2-}		28.5	8.9
Specific activity:			
Cs-134	Bq/l	$2.33 \cdot 10^6$	$4.06 \cdot 10^7$
Cs-137		$4.46 \cdot 10^6$	$2.41 \cdot 10^5$
Co-60		$2.01 \cdot 10^5$	$2.35 \cdot 10^4$
Na-24		$8.31 \cdot 10^4$	-

Ratio of the basic radioactive elements in wastes is as follows: Cs-137 and Cs-134 - about 90 %, Co-60 - about 10%.

As binders in experiment, portland-cement, slag-portland-cement, alkaline binders (geocements) of various compounds were used. To obtain geocements, ground, granulated metallurgical slags with the basify modulus $M_o = 1.1$ were used. To enhance the isolating effect, modifying additives of natural origin have been introduced into geocements. Their action is based on their own sorption properties and is also provided by their direct participation in the synthesis of zeolite-like neo-hydration products.

The cementing, that is, the inclusion of liquid radioactive industrial wastes into mineral cementitious materials, is recognized now by specialists worldwide as the most perspective conditioning technology in terms of simplicity and cost.

However, when it is applied to wastes with a salt content up to 200 g/l, only 5 to 7 % by mass of them are bound within the compound produced. That is why to increase a salt content in the compounds formed the use was made of an evaporation technique applied to the concentrate in laboratory conditions, and also the water to cement ratio was increased up to 0.6. As a result, the Khmelnitskiy nuclear power plant's radioactive waste dry residue concentrations increased to 49-51 % by mass, and those at Chernobyl up to 47-65 % by mass. These degrees of evaporation correspond to salt concentrations for the Khmelnitskiy nuclear power plant - about 600-700 g/l and for the Chernobyl nuclear power plant - 646-991.9 g/l.

The evaporated concentrate was mixed with other constituents to produce a flowable mixture and prism- specimens (20 × 20 × 80 mm) were cast, and after 1 day were cured under normal conditions (T = 20°C, RH = 95-100%) before testing

To determine the degree of reliability of isolation, the leach rate was examined. After 56 days starting from the preparation the specimens have been tested for compliance to IAEA and ISO 6961-1982 standards (E).

RESULTS

As it follows from the data obtained, strength of the specimens made with geocements, in which the concentrate from the Khmelnitskiy nuclear power plant was used as a geocement constituent, was at an age of 28 days (this age corresponds to a cement strength class) - 12.7 and 17.3 N/mm^2, thus exceeding considerably the minimum allowable value of 10 N/mm^2. Specimens made with Portland cements (PC) had a strength of 9 N/mm^2, and slag-Portland-cements (SPC) had a strength of 11 N/mm^2. The salt content in all the resulting compounds was about 26.5 % by mass.

Strength of the specimens made with geocements, in which the concentrate from the Chernobyl nuclear power plant was used as a geocement constituent, was at an age of 28 days 11.4 and 18.8 N/mm^2 at a salt content in the resulted compounds - 28.6 and 21.1% by mass, respectively. For specimens made with Portland cements (PC) - 5 N/mm^2, slag-Portland-cements (SPC) - 6.2 N/mm^2. The salt content in all the resulted compounds was about 13 % by mass.

According to conclusions made by the commission of the Khmelnitsk nuclear power plant, the leach rate of Cs-137 from the resulting compounds made with geocements after 28 days was 8.3·E-05 - 9.43·E-05 g/sq cm day, being considerable lower the allowable limit - 1·E-03 g/sq cm day, whereas those from the resulted compounds made with PC were 1.56·E-03 and SPC - 2.47·E-03 g/sq cm day, respectively.

Test results of leach rate of radionuclides held at the Khmelnitskiy and Chernobyl nuclear power plants under the IAEA testing procedure are shown in Figure 1 and Table 2.

Worthy mentioning is the fact that in the resulted compounds made with geocements the leach rate stabilized by 28 days starting from the moment of its determination, whereas in those made with PC and SPC - by 98-154 days.

When using the concentrate from the Chernobyl nuclear power plant, the leach rate of Cs-137 from the resulted compound made with geocement was by 1 day - 2.69·E-03 - 9.62·E-04 g/sq cm day, by 10 days age from the time of determination declined to 2.14·E-04 and 8.83·E-05 g/sq cm day. In the compounds made with portland cement and slag portland cement - the stabilization of leaching took place by 28-35 days from the time of determination and was 4.48 E-03 - 5.95·E-03 g/cm^2 per day.

Table 2 The results of tests of cement compounds with liquid radioactive industrial wastes of atomic power stations

INDICES	KHMELNITSKIY NPP, THE REACTOR WWER				CHERNOBYL NPP, THE REACTOR RBMK			
SAMPLE CODE	H-1	H-2	K-2	K-9	H-1	H-2	K-1	K-2
CONTENT OF SALTS, %	26.5	26.5	26.5	26.5	13	13	28.6	21.1
ASPECT OF BINDERS	PC	SPC	Geocement		PC	SPC	Geocement	
COMPRESSIVE STRENGTH, N/mm^2, AT AN AGE OF, DAYS								
28	9.0	11.0	12.7	17.3	5.0	6.2	11.4	18.8
56	11.3	11.2	14.5	19.3	5.7	6.8	14.7	23.4
LEACH RATE CS^{137}, g/cm^2 DAY, AT AN AGE OF DAYS:								
1	$2.7\cdot10^{-2}$	$3.7\cdot10^{-2}$	$3.7\cdot10^{-4}$	$7.6\cdot10^{-4}$	$2.4\cdot10^{-1}$	$1.5\cdot10^{-1}$	$2.6\cdot10^{-3}$	$9.6\cdot10^{-4}$
3	$1.1\cdot10^{-2}$	$9.9\cdot10^{-3}$	$2.7\cdot10^{-4}$	$4.5\cdot10^{-4}$	$6.7\cdot10^{-2}$	$4.8\cdot10^{-2}$	$9.2\cdot10^{-4}$	$4.1\cdot10^{-4}$
7	$4.5\cdot10^{-3}$	$4.1\cdot10^{-3}$	$2.3\cdot10^{-4}$	$2.8\cdot10^{-4}$	$3.0\cdot10^{-2}$	$1.9\cdot10^{-2}$	$3.7\cdot10^{-4}$	$1.7\cdot10^{-4}$
10	$3.6\cdot10^{-3}$	$3.5\cdot10^{-3}$	$2.2\cdot10^{-4}$	$2.5\cdot10^{-4}$	$2.0\cdot10^{-2}$	$1,1\cdot10^{-2}$	$2.7\cdot10^{-4}$	$1.4\cdot10^{-4}$
14	$2.8\cdot10^{-3}$	$2.5\cdot10^{-3}$	$1.7\cdot10^{-4}$	$2.2\cdot10^{-4}$	$1.2\cdot10^{-2}$	$7.1\cdot10^{-3}$	$2.2\cdot10^{-4}$	$1.1\cdot10^{-4}$
21	$2.6\cdot10^{-3}$	$2.6\cdot10^{-3}$	$1.6\cdot10^{-4}$	$1.5\cdot10^{-4}$	$9.0\cdot10^{-3}$	$5.7\cdot10^{-3}$	$2.1\cdot10^{-4}$	$8.6\cdot10^{-5}$
28	$2.5\cdot10^{-3}$	$3.2\cdot10^{-3}$	$8.3\cdot10^{-5}$	$9.4\cdot10^{-5}$	$6.0\cdot10^{-3}$	$4.5\cdot10^{-3}$	$1.5\cdot10^{-4}$	$7.6\cdot10^{-5}$
35	-	-	-	-	$4.7\cdot10^{-3}$	$3.1\cdot10^{-3}$	$1.2\cdot10^{-4}$	$5.3\cdot10^{-5}$
59	$1.4\cdot10^{-3}$	$1.4\cdot10^{-3}$	$7.8\cdot10^{-5}$	$6.1\cdot10^{-5}$	-	-	-	-
98	$7.1\cdot10^{-4}$	$7.5\cdot10^{-4}$	$4.1\cdot10^{-5}$	$3.5\cdot10^{-5}$	-	-	-	-
126	$5.6\cdot10^{-4}$	$5.0\cdot10^{-4}$	$3.9\cdot10^{-5}$	$3.1\cdot10^{-5}$	-	-	-	-
154	$4.6\cdot10^{-4}$	$4.5\cdot10^{-4}$	$3.6\cdot10^{-5}$	$2.4\cdot10^{-5}$	-	-	-	-
193	$4.2\cdot10^{-4}$	$4.7\cdot10^{-4}$	$3.1\cdot10^{-5}$	$1.9\cdot10^{-5}$	-	-	-	-
215	$5.2\cdot10^{-4}$	$4.2\cdot10^{-4}$	$2.2\cdot10^{-5}$	$1.4\cdot10^{-5}$	-	-	-	-

a)

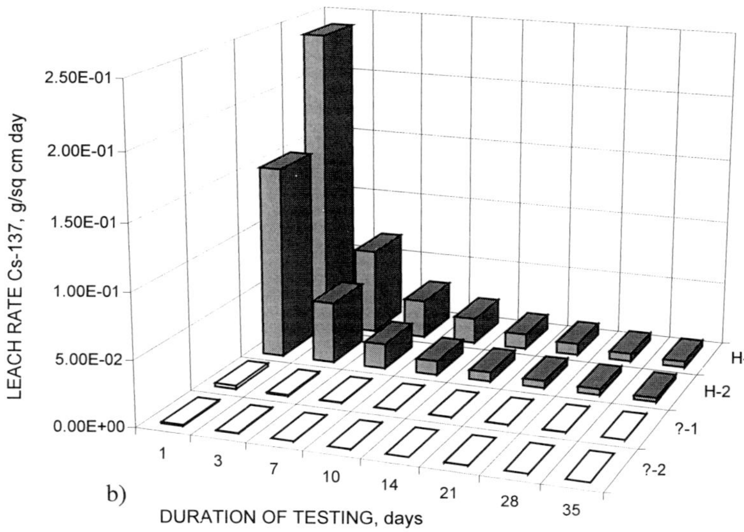

b)

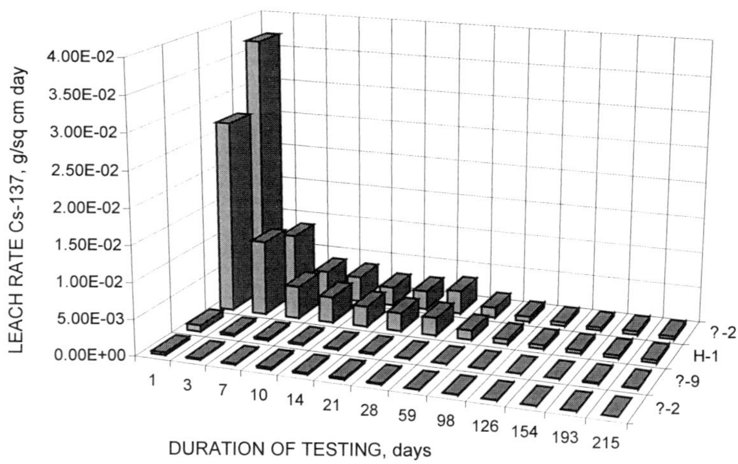

Figure 1 Leach rate of Cs-137 from cement compound into water

a) Liquid radioactive wastes, Chernobyl NNP, reactor RBMK
b) Liquid radioactive wastes, Khmelnitskiy NNP, reactor WWER
K-1, K-2, K-9 – geocement, a matrix
H-1, H-2 – portland cement and slag portland cement, a matrix

CONCLUSIONS

1. The developed approach to immobilization of liquid radioactive wastes demonstrated hat the use of a new class of mineral hydraulic cementitious materials - *geocements* - gives rise to environmentally responsible long-term storage by providing a strong chemical fixation of radionuclides within the mineral-like cement stone matrix.

2. The technology is advantageous in terms of simplicity and cost. These advantages come from low cost of geocement constituents, since a wide range of industrial wastes and by-products as well as widely available natural raw materials are used, and the facilities that are commonly used now can be also involved. The results of a feasibility study showed that the manufacturing cost of the compounds is about 15-30 % lower (comparison based on raw materials cost).

3. Of great advantage in terms of minimizing environmental risk and cost of maintenance, is the greater durability of the compounds made with geocements which can be predicted. This prediction is supported by knowledge of the zeolite-like formations formed in the resulted compounds. This is because zeolites are recognized among natural rocks as having excellent stability and strength properties.

4. Following the successful laboratory trial, the effect of geocements on industrial equipment was checked. In addition, a technological line for cementation of liquid and rigid radioactive wastes with use of geocements has been assembled and successfully operated at the Chernobyl NPP.

REFERENCES

1. GLUKHOVSKY V D. Soil Silicate Articles and Constructions. Budivelnik Publish., Kiev, USSR, 1967, 154 pp.

2. KRIVENKO P V. Acid Resistant Materials Based on Alkaline Alumosilicate Binders. Cand.Tech.Sc.Degree Thesis, Civil Engineering Institute, Kiev, USSR, 1971, 24 pp.

3. SKURCHINSKAYA J V. Synthesis of Analogues of Natural Minerals with the Aim of Producing of Artificial Stone, Cand.Tech.Sc.Degree Thesis, Civil Engineering Institute, Kiev, USSR, 1973, 24 pp

4. GLUKHOVSKY V D. Soil Silicate: Their Properties, Production Technology, Field of Application, Doct. Tech.Sc.Degree Thesis, Civil Engineering Institute, Kiev, USSR, 1962, 48 pp.

5. GLUKHOVSKY V D. ET AL: Alkaline and Alkaline-Earth Alumosilicate Hydraulic Binders and Concretes, Vischa Shcola Publ., Kiev, USSR, 1979, 232 pp.

6. KRIVENKO P V. Synthesis of Binders with Special Properties in a Sistem Me_2O-MeO-Me_2O_3-SiO_2-H_2O, Doct. Tech.Sc.Degree Thesis, Civil Engineering Institute, Kiev, USSR, 1986, 48 pp.

7. KRIVENKO P V.: Alkaline Cements, Proc.9-th International Congress on the Chemistry of Cement, New Delhi, India, 1992, vjl.4, pp.482-488.

8. KRIVENKO P V ET AL: Usage of Geocements of Radioactive Waste Immobilisation, Proc.XV International Mendeleev`s Symposium, Obninsk, Russia, 1993, p.p. 89-90.

9. GLUKHOVSKY V.D. ET AL: Composite of Solidication of Metal Carbonate Containing Radioactive Waste, USSR, Author`s Certificate (Patent) № 1471571, priority of 17.03.87 (unpublished).

10. GLUKHOVSKY V D, ET AL : Processing Technique for Alkali Metal Containing Radioactive Waste, USSR, Author`s Certificate (Patent) № 1625252, priority of 31.03.89 (unpublished).

INNOVATIVE FORMS OF ENVIRONMENTALLY-SENSITIVE PRECAST CONCRETE PAVING

B Shackel

University of New South Wales

Australia

ABSTRACT. This paper describes the laboratory testing of new forms of precast concrete pavement comprising both permeable concrete block paving and precast concrete slabs linked at their corners by plastic connectors. These forms of paving are designed for manufacture by simple plant and to minimise adverse environmental impacts whilst using unskilled labour. The paper begins by assessing the environmental needs that precast paving can address. Detailed laboratory and field evaluations of the paving systems are then reported. The experimental data show that these new forms of paving are viable for traffic loads and numbers of repetitions similar to those likely to be applied by both trucks and medium size aircraft whilst, at the same time, providing substantial and quantifiable environmental benefits.

Keywords: Concrete, Environment, Infiltration, Stiffness, Pavement, Paving, Permeable, Precast.

Dr Brian Shackel is an Associate Professor in Civil and Environmental Engineering at the University of New South Wales, Sydney, Australia. He is a specialist in pavement engineering and is the author of more than 100 research papers and a book on concrete block paving currently available in English German and Japanese editions. He has been invited to lecture on pavements in more than 25 countries worldwide.

INTRODUCTION

To qualify as being environmentally sensitive, paving should be:

1. Simple to manufacture.
2. Simple to lay without the need for expensive or complicated construction plant. Ideally, it should be capable of being installed by unskillled labour.
3. Economical both to lay and to maintain and in the use of resources such as base and sub-base materials
4. Low in adverse environmental impacts during paver manufacture, construction or in-service.

Many of these issues can be addressed by using precast concrete paving which is permeable and which minimises the use of base and sub-base materials. First, permeable eco-paving is discussed. A new form of paving known as Enviropave is then described. which eliminates the need for expensive pavement materials by making more effective structural use of natural subgrades than has been possible hitherto.

Both forms of paving comprise small precast concrete pavers. laid on a granular bedding course. For permeable paving the bedding course overlays a basecourse and (optionally) a sub-base over the subgrade and therefore has a sub-structure similar to that of a conventional flexible or concrete block pavement. By contrast, the Enviropave system is laid directly on a sand subgrade which fulfills the roles of bedding, base and sub-base.

PERMEABLE PAVING

The essential basis of eco-paving is that the pavement should be permeable and that water infiltration should be actively encouraged. The environmental benefits of permeable paving are:

a) Eco-paving reduces the amount of rainfall runoff from pavement surfaces and, thereby, eliminates or minimises the extent of the stormwater drainage system required.
b) Eco-paving reduces the size or need for rainwater retention facilities by using the pavement itself for retention. This maximises land use.
c) Eco-paving reduces or avoids downstream flooding.
d) Eco-paving helps recharge and maintain aquifers and the natural groundwater.
e) Eco-paving helps trap pollutants that would otherwise contaminate groundwater or drainage systems.
f) Eco-paving assists in the biological decomposition of hydrocarbon contaminants from vehicle depositions.

The derivation of a typical eco-paver from conventional is shown schematically in Figure 1. Essentially, eco-pavers are manufactured by modifying well-established concrete paver shapes so that, once layed, openings are provided at intervals along the joints between the pavers. These are filled with gravel to act as vertical drains through the pavement. Because eco-pavers are based on well-proven conventional paver shapes and are intended to be installed on a fully engineered pavement sub-structure, they provide an alternative to conventional segmental paving provided that it can be demonstrated that they can combine rapid water infiltration with good load carrying capacity.

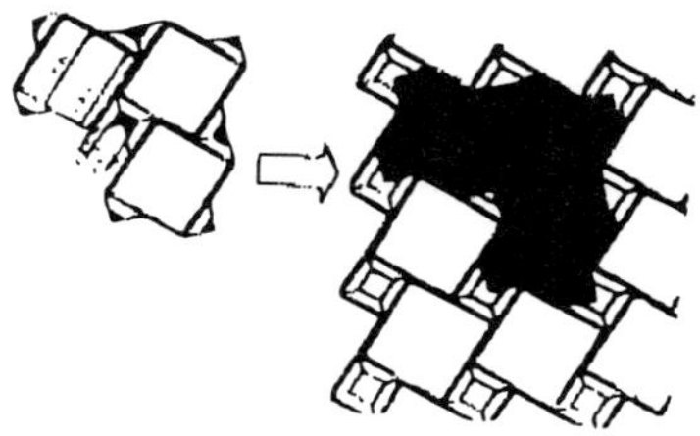

Figure 1 Evolution of permeable Uni-Ecoloc from conventional Uni-Coloc pavers

Infiltration Tests

The infiltration tests evaluated normally constructed areas of paving in a rainfall simulator and h ave been described in detail elsewhere (1). Water draining through the paving and bedding layer and runoff from the pavement surface could be collected and measured separately. A range of 80mm thick eco-pavers was studied including that shown in Figure 1.

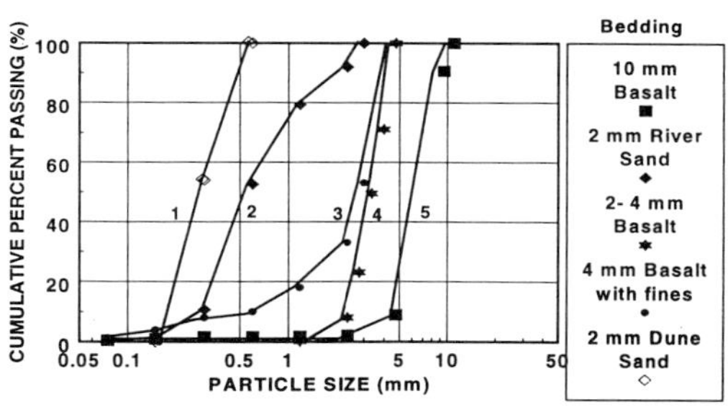

Figure 2 Grading of the materials used in the experiments

As shown in Figure 2 (curves 2 to 5) the bedding materials studied ranged from a 2mm river sand to a 10mm aggregate. In the case of the10mm aggregate (curve 5 in Figure 2), relatively little material penetrated the joints between the pavers during construction. For this reason, a further test was conducted in which the 10mm aggregate was used as the bedding and jointing material, but the 2mm river sand (curve 2) was brushed into the joints after compaction. This not only filled the joints but effectively blinded the drainage openings.

Typical results of infiltration tests pavements are given in Figure 3. This figure shows that the pavements utilising the coarse gravel accepted the full intensity of the simulated rainfall without surface runoff. However, there was about a 50% reduction in water infiltration in the pavements laid using only sand or where sand was used to blind the surface. Figure 3 shows that there was little or no benefit in using gravel in the drainage openings if it is to be capped with sand. For this reason, in all subsequent tests, only a single material was used as the bedding, jointing and drainage medium and no further infiltration studies were conducted using conventional 2mm bedding sands. Moreover, because of the lack of joint filling fractions in the 10mm gravel, further study was restricted to materials 3, 4 and 5 in Figure 2 ie. to materials having particle sizes predominantly within the range 2mm to 5mm.

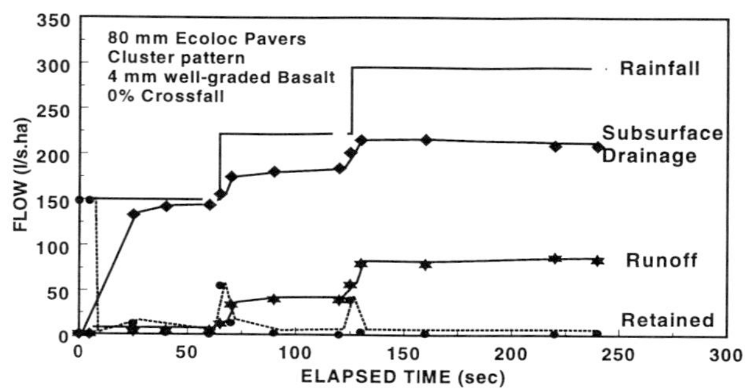

Figure 3 Typical infiltration test data for permeable eco-paving

In these tests, the principal determinant of performance was the grading of the drainage and jointing materials. For all three of the 5mm gravels, rainfall intensities of 150 l/s.ha or more could be wholly accepted by the eco-paving surface. The highest rates of water acceptance were associated with the relatively uniform 2-5mm aggregate. Here, rainfall intensities of about 600 l/s.ha could be wholly accepted by the paving. Of the two materials containing fines, the 5mm gravel with fines was associated with higher levels of infiltration than the 4mm well-graded gravel. In general, uniform materials without fines showed better drainage characteristics than where fines were present and infiltration decreased with increase in the fines content.

Structural Tests

A simple test procedure was used which allowed the pavers to be tested in isolation from the pavement sub-structure. Details have been given elsewhere [2]. Despite its simplicity the test has been shown to yield the same ranking of the effects of paver shape etc. as that given by more complicated and expensive accelerated trafficking tests [2,3]. The test involved laying and compacting the pavers in a steel frame placed upon the basecourse. Following compaction, the frame, together with the intact paver mat, could be lifted off the base. Under the self-weight of the pavers the centre of the mat of pavers sagged sufficiently (2-8mm) to induce arching and wedging between the individual pavers, and between the pavers and the test frame. This caused the pavers to achieve structural integrity and to act like an articulated slab. Despite the lack of any support, the pavers were then capable of supporting significant vertical load. The unsupported mats of pavers could be tested like a slab beneath a jack having a capacity of 25 kN. This applied a central load to the mats via a rigid circular plate having a diameter of 150mm. The deflections of the mats were measured so that the equivalent quasi-elastic modulus of the paver mats could be measured.

It was noted earlier than eco-pavers are modified from conventional pavers. Figure 4 shows a comparison bewteen the mat modulus measured for two common eco-pavers (A and B) and the conventional paver from which A was derived. It may be seen that the eco-pavers exhibited similar moduli to than the conventional pavers. Overall, however, the provision of the drainage openings along the joint lines and the use of coarser bedding materials than those used customarily did not significantly alter the ability of eco-paving to carry load.

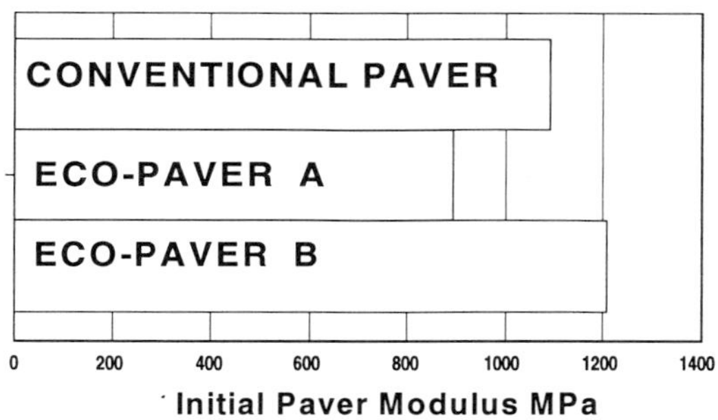

Figure 4 Comparison of the moduli of conventional and eco-pavers

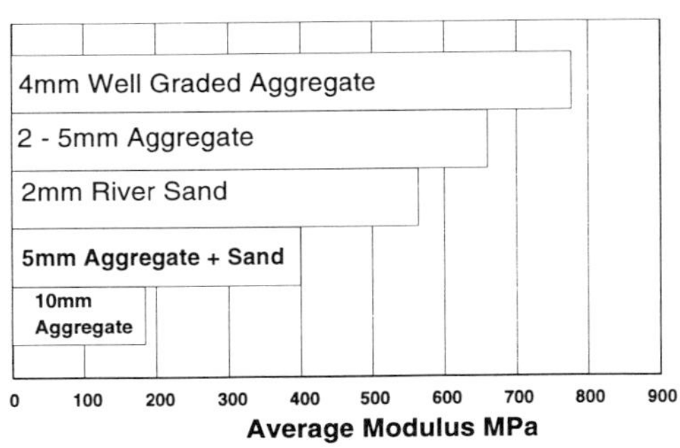

Figure 5 Effects of the bedding and jointing materials on the moduli of eco-paving

The effects of varying the bedding and jointing materials on the mat stiffness are shown in Figure 5. The best levels of performance (ie. the highest moduli) tended to be associated with those pavements incorporating jointing materials capable of penetrating and filling the 2-4mm wide joints between the pavers. In this respect, the pavements laid with the 10mm gravel exhibited low values of moduli irrespective of whether or not the joints were blinded with sand. Overall, it can be concluded that the material used as the bedding, jointing and drainage material in eco-paving should include a sufficient proportion of material size capable of penetrating the joints ie no bigger than 4mm. However, as shown in Figure 5, the coarse aggregates were associated with higher moduli than the conventional 2mm bedding sand. Overall, in common with other types of concrete segmental paving [3] the eco-paving was found to be structurally efficient.

ENVIROPAVE

In many other parts of the world, there are large areas of sand terrain. Often this terrain is remote from large population or manufacturing centres and poses logistical problems whenever pavements must be constructed. A need therefore exists for paving systems capable of being layed in remote areas without normal construction plant and requiring only minimal importation of materials. The evaluation of a new environmentally friendly concrete paving system developed to address these needs is now described. The system is known as Enviropave.

As shown schematically in Figure 6, Enviropave consists of concrete slabs which are 390 mm by 290 mm in plan and just 90 mm thick made using normal concrete precasting plant. When laid, the slabs are separated by joints that are 10 to 12mm wide. To achieve structural continuity plastic connectors are used to link the slabs. These connectors consist of mats made from recycled plastic which are 290 mm square in plan and which support 4 vertical spigots. The slabs themselves are manufactured with tapered vertical holes in each corner. These enable the pavers to hook on to the spigots. Plastic wedging cones are then forced or screwed into the spigots to lock the slabs together and, thereby, to eliminate horizontal creep.

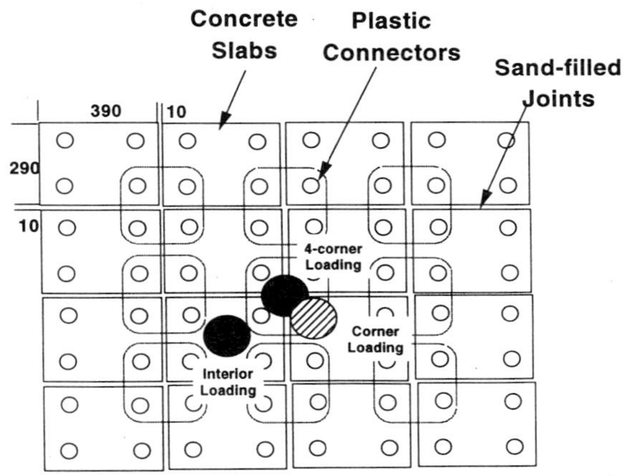

Figure 6 The Enviropave system

Enviropave is designed to facilitate rapid, non-invasive and easy construction of roads and runways over sand terrain. The system eliminates the need for excavation and the use of imported road base. Rather, the precast concrete slabs form a continuous but articulated concrete pavement. This articulated mat uses the *in-situ* sand subgrade as the base. The use of small slabs reduces the tensile stresses induced by traffic loading and permits the use of relatively thin, light and easily transported slabs. Environmental impacts and cost are minimised because the paving does not require a base or sub-base. The system can be easily hand laid over a graded surface using unskilled labour without the need for large or heavy construction equipment.

The system is suited to the construction of access roads over sand terrain to carry heavy equipment. Other applications include parking areas, boat ramps, sand stabilisation, causeways, embankments, country saleyards and feed lots. However, in Australia, the principal engineering challenge was to provide all-weather runways for aircraft access to remote outback areas.

Water infiltration and structural repeated plate loading trials have been conducted on Enviropave laid on a windblown dune sand subgrade having gradation 1 in Figure 2. The infiltration tests showed that Enviropave could accept water flows of approximately 150 l/s.ha. This was much lower than can typically be achieved in eco-pavers and was primarily a reflection of the need to use only sand to bed and joint the pavers but also of the small area of joints and openings able to act as drains relative to the pavement surface area.

In the structural tests, the sand was tested in an air-dry condition (moisture content = 1.8 %). The tests have been described in detail elsewhere [4] and involved measuring surface deformations and deflections and sub-surface stress distributions under loads up to 57 kN applied at a contact (tyre) pressure of 1.4 MPa with upto 20 000 stress repetitions at 0.2 Hz. The tests explored the effects of load positioning and magnitude, the contribution of the plastic connectors to the pavement performance and the role of compaction.

Effect of Load Position and Magnitude

As shown in Figure 6, the pavements were tested with the load applied at the centre of a slab (interior loading), at the junction between four slabs (ie centred over the plastic connector) or entirely within the corner of one slab. Both the permanent deformations and resilient deflections were higher for interior loading than for corner loading when the slabs were layed without compaction. For these uncompacted pavements, the deformations and deflections were much higher than could normally be tolerated in trafficked pavements. However, for slabs on compacted subgrades the placement of the load had only a very small effect, Here the permanent deformations were slightly higher when the load was placed within the corner of a single slab than when the load was placed at the interior of a slab or over the junction between four slabs.

For pavements on compacted subgrades the effects of increasing the applied load were studied. The tests were terminated once the deformations exceeded 20 mm because this represented the level at which the pavements would be regarded as becoming unserviceable. As shown in Figure 7, for loads of 37 and 57 kN, the effects of load placement were more critical than for the 19 kN load and the accumulation of deformation was more rapid.

At a load of 37 kN, the pavement could withstand approximately 10 000 repetitions of load before reaching the limit of serviceability. For the 57 kN load this dropped to just 200 load repetitions.

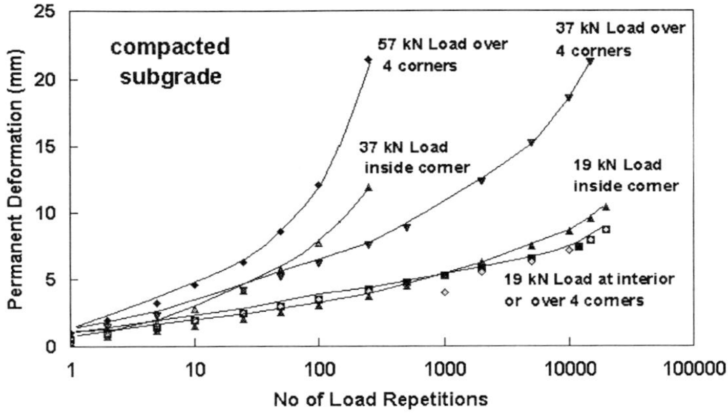

Figure 7 Effects of load magnitude and placement on Enviropave

Effect of Subgrade Compaction

Trial pavements installed without compaction exhibited large deformations and deflections. The effects of compaction were, therefore, studied. Compaction was achieved by applying a light vibratory plate compactor to the surface of the slabs after they were layed. No attempt was made to compact the loose sand before slab installation. This was done to make construction of the pavements as simple as possible. Ten cycles of plate compaction produced total vertical compressions of between 50 and 60 mm despite the small size and weight of the compactor. This reduced the deformations of the pavements under repeated loading. from about 90 mm for the uncompacted pavements at 10 000 load repetitions to just 8 mm at 20 000 repetitions for the compacted pavements. At the same time the resilient (elastic) deflections were reduced from between 2 and 4 mm to approximately 0.3 mm

Effects of the Connectors

In previous paving systems based on the use of small slabs, connection between the units has usually been by means of steel bolts or dowells [eg 5]. These systems attempt to achieve load transference from one slab to the next. By contrast, the plastic connectors used in the Enviropave system have low rigidity and would not be expected to provide much load transference. Furthermore, for uncompacted pavements, there were significant deformations, deflections and slab rotations under load and the contribution of the plastic connectors to the structural performance of the pavements was not clear. For this reason, a pavement was installed without connectors and was subjected to loading distributed over the corners of four adjacent slabs. This pavement failed during the first load application with the deformation exceeding 50 mm. This demonstrated that the connectors did, in fact, significantly contribute to structural performance under load ie their role was not just to locate the slabs relative to one another. In this respect, it is important to note that, as shown in Figure 6, the connectors were 290 mm square bearing directly on the subgrade. This meant that corner loads on the slabs were more widely distributed than had the slabs been placed directly on the subgrade. Moreover, there was no significant faulting or stepping between adjacent slabs. This indicated that load transference was adequate for the proposed traffic load levels.

CONCLUSIONS

The experimental work showed that it is feasible to use precast concrete paving to achieve the environmental benefits of both accepting substantial water penetration and minimising the use of materials in the pavement sub-structure. Specifically it was found that:

1. It was possible to reconcile the requirements of obtaining good water infiltration with adequate structural capacity under laboratory conditions by using permeable eco-pavers.

2. Both water infiltration and stiffness depended on the gradation of the materials used to bed the eco-pavers and to fill the joints and vertical drains.

3. The best structural performance and water penetration was achieved using a 2-4mm graded aggregate as bedding and jointing material. Here infiltrations of up to about 600 l/s.ha were measured. This is adequate for most practical drainage requirements.

4. There were only marginal differences between the stiffnesses of the eco-pavers studied and the conventional pavers from which their profiles had been derived.

5. Increase in the fines present in the jointing and drainage material led to a reduction in the ability of the eco-pavements to accept rainfall without any commensurate structural benefit.

6. It was possible to avoid the use of base and sub-base materials on sandy terrain by adopting the Enviropave system consisting of small precast concrete rafts linked by plastic connectors and laid directly on the subgrade.

7. The Enviropave system was capable of accepting concentrated repeated loads up to 57 kN whilst maintaining acceptable performance levels in terms of deformation and deflection.

8. Much of the structural integrity of the EnviroPave system derived from the plastic connectors despite their low structural rigidity.

9. Satisfactory performance of the trial pavements under repeated loading could not be achieved where the slabs were laid directly on loose sand without compaction.

10. Placing the slabs on loose sand and then compacting the pavements with a plate compactor significantly reduced the deformations and deflections induced by repetitive loading

ACKNOWLEDGEMENTS

The equipment and work described in part of this paper was funded by Besser Pioneer and Viscount Plastics, Queensland. The assistance of these companies is gratefully acknowledged.

REFERENCES

1. SHACKEL B. Water Penetration and Structural Evaluations of Permeable Eco-paving. Betonwerk und Fertigteil-technik Vol 63, No3, March 1997, pp110-119

2. SHACKEL, B., WINARNO AND O'KEEFFE, L. Concrete Block Paving Tested as Articulated Slabs. Proc. 5 Int. Conf. on Concrete Pavement Design and Rehabilitation, Purdue University 1993

3. SHACKEL, B., Design and Construction of Interlocking Concrete Block Pavements London, Elsevier Applied Science, 1990, 229pp.

4. SHACKEL B, O'KEEFFE L J, GWYNNE P. W AND ARISDIANTO I. Environmentally Sensitive Articulated Concrete Pavement. Proc. 6th Int. Conf. On Concrete Pavements, Indianapolis, 1997

5. CENTRAL ROAD RESEARCH INSTITUTE, INDIA The development of special purpose paving blocks in India, Proc 2nd Int Conf. On Concrete Block Paving, Delft . 1984, pp 387-344.

APPLICATION OF THE RILEM METHODOLOGY FOR THE DURABILITY DESIGN OF NEW CONCRETE SILO RESERVOIRS RETAINING STRUCTURES

L M Poukhonto

Moscow State University of Civil Engineering

Russia

ABSTRACT. The increasing amount of damage during the use of concrete silo, bunkers and retaining structures demands further perfection of durability design. This study is partly based on the RILEM TC 130 -CSL approach and a lifetime safety factor method is developed. On the other hand some of the authors' results concerning damage cumulation analyses, parameters of stationary cyclic actions and the residual life span assessment are also given. A theoretical degradation model for assessing the corrosion of reinforced steel in structural concrete, considering several deterioration mechanisms in conjunction with low-cyclic load is presented . The paper briefly shows the results of numerical simulation, based on the application of non-linear finite element method (FEM) and gives an example of the structural durability design procedure for the case of a structural grain silo.

Keywords: Concrete silo, Bunkers, Retaining structures, Deterioration, Low-cyclic loading, Damages cumulation, Service life, Prediction, Deterioration model, Lifetime safety factor method, Non-linear finite element method.

Dr Leonid M Poukhonto is an Associate Professor of the Department of Reinforced Concrete Structures, Moscow State University of Civil Engineering. His research interests include: deterioration and damage of concrete structures, deteriorate assessment, inelastic behaviour, computer design, service - life prediction and durability of concrete structures: silos, bunkers and reservoirs and probabilistic characteristics of low-cyclic loading, cumulation damage analysis and evaluation. For the past 10 years he has been involved in research on durability and service life prediction of reinforced concrete structures under low-cyclic load . His D Sc. (Eng) thesis is now in the closing stages.

INTRODUCTION

Concrete constructions of silo, bunkers and reservoirs, water towers, retaining walls in many countries shows in a number of cases an increasing degree of damages and deterioration within a short of span 1 - 3 to 15 years. As a new illustration it is possible to result an example of unexpected destruction soya meal silo, capacity 5400 t , service life of which before failure has made 13 years [1 - 3].

This topic reviews the main aspects of the authors research and reflects the point of view on durability on an international basis, including RILEM, EVROCOD and ASIA MODEL CODE recommendations [4]. The paper deals with the application of semi - stochastic and non -linear finite element methods to develop durability design of concrete structures mentioned .

DEVELOPMENT OF DURABILITY PREDICTING METHODS

In practice deterioration of spatial concrete structures is seldom due to a single cause. The durability model for service life prediction with regard to reinforcement corrosion in conjunction with low cyclic load has been developed, using same holistic approach by P. K. Mehta (1994) [5], see also [6-8]. The influence effect of the several degradation processes has been considered in J. Clifton (1993), S. Alekseev (1993), A .Sarja, E. Vesikary and Fagerlund (1994),V. Almazov (1995), E. Guzeev (1995), Jarewal, Kumar, Trikhna (1997), V.Bondarenko (1998) and others.

Solving the durability problem of existing and new concrete structures of the storage constructions, sources published during 1980 - 1998 have been analysed by the author [6,7]. The purpose of this part of study consisted in the review recent science - based durability assessment techniques and predicting methods and their analysis to establish the most acceptable general methods of the material and structural durability assessment of concrete structures . Results of this study showed that the methods used for the service life predicting of reinforced concrete structures are the following :

- Deduction from performance of similar reinforced concrete structures
- Accelerating testing (mostly materials of structures)
- Mathematical modelling based on study of degradation processes
- Application of reliability and stochastic concepts
- Fracture mechanic approach
- Linear and non-linear analysis of reinforced concrete structures by the finite element method.

PRACTICAL DURABILITY ASSESSMENT

Service life analyses aims at creating the performance over time functions describing how the essential properties of the product vary with time [4, 7, 9]. A structural load of grain silo, that varies in time, has been modelled, using stochastic analyses for nonstationaty time series of grain loading [6].

The RILEM lifetime safety factor method has been developed for durability design of R.C. element of construction mentioned based on the theory of safety and reliability (semi - stochastic method). The determination by stochastic methods and the meaning of life - time safety factors for structural durability design of concrete constructions, safety margin requirement seems to be very important taken into account this approach. The design service life is determined as [4]:

$$t_d = \gamma_t \, t_g \tag{1}$$

where t_d = the design service life ; γ_t = the lifetime safety factor, determined with a lognormally distributed service life function ; t_g = the target service life.

The example of the design procedure was executed to find out the durability parameters of undersilo columns subjected to reinforcement corrosion in conjunction with low-cyclic loading. The durability design procedure contains : i) specification of the target service life and design service life; ii) identification of durability factors and selection of a appropriate durability model; iii) calculation of durability parameters (depth of deterioration, concrete cover, reduction of safety margin) using available model. The undersilo columns to be dimensioned with square cross - section 500 x 500 mm for the loads: F_g= 971 kN (dead load) ; F_p=2818 kN (variable grain load) .The following parameters were received : maximum depth of cover C_{min}= 67 mm ; the relative reduction of safety margin $m=(\theta_0-\theta_m)/\theta_0$=0.47 .

Following Andrade, Wang and other researchers [10 - 12] a two-phase model of corrosion is considered as the first approximation, consisting of periods of sequential penetration of front of corrosion in a cover of concrete and of distribution of corrosion on volume, occupied by the steel bar. The model of moving front of penetration of aggressive in concrete and development of uniform corrosion of a steel rod is given on Figures 1 a and 1 b .

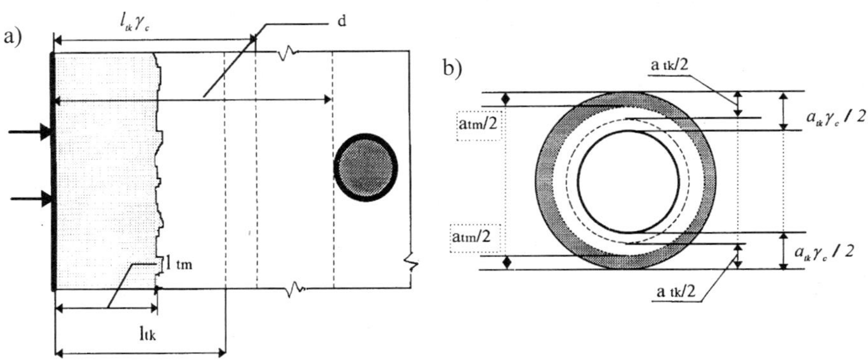

Figure 1 a) to determination of characteristic value of depth of penetration of aggressive environment : d - depth of concrete cover ;l_{tm} - mean value of penetration depth ;l_{tk} - characteristic value of the mean penetration depth ; b) to determination of characteristic value of thickness of a damaged layer of a rod at uniform corrosion : a_{tm} - mean value of thickness of damaged bar layer ; a_{tk} - characteristic significance of depth of a damaged layer of the reinforcement

The guaranteed level of size of a concrete cover d_m is expressed in a form :

$$d_m = d_c + ks \tag{2}$$

where $\dot{d}_c$ is the normalized significance of depth of the cover; s is the standard deviation; k is constant based on the characteristic of safety, connected through normal distribution with a level that is lower as cover defects meet. k varies between 1.28 - 2.33 at fluctuations of defects from 1% up to 10 %

Depth of a layer of corrosion of the reinforcement a_{tk} is expressed as:

$$a_{tk} = v_{c,t,k} c_{cor} t_{cor} \tag{3}$$

$v_{c,t,k}$ is the characteristic significance of average speed of corrosion of the fixture by results of tests, c_{cor} is a factor taking into account a difference between experienced samples and by a real structures ; t_{cor} is the duration of corrosion of the reinforcement, years.

For the structures, for which corrosion of the steel reinforcement is not allowed, the criterion of durability can be recorded as :

$$l_{tk} \le d_{tk} \tag{4}$$

where l_{tk} is the characteristic value of the mean depth of penetration ; d_{tk} is the characteristic value of the mean depth of concrete cover, and z is the reserve of safety.

$$z = d_{tk} - l_{tk} \tag{5}$$

NON-LINEAR ANALYSIS OF REINFORCED CONCRETE CYLINDRICAL WALL OF SILO STRUCTURE BY THE FINITE ELEMENT METHOD

In repeatedly statically indeterminate structures the contribution of separate constructive elements to bearing ability of a structure is as a whole variable. Damage or complete destruction of one or more elements cause loss of the bearing capacity, but do not result in complete loss in connection with redistribution of inner forces . Properties of the concrete and reinforced steel and thickness of the silo wall change from the effects of the environment with time, and the combination of corrosion and low - cyclic stresses accelerates this process.

The influence of the limited penetration of aggressive environement on the stress-strain state in the silo walls of the cylindrical form are studied in the given section with the help of numerical modelling.

Using durability design procedure the account of corrosion wear is reduced to the research of a cylindrical shell with the thickness variable in time. The law of change of thickness for average zone on height of an shell accepted in a form:

$$h(t) = h_0 - \varphi(t) \tag{6}$$

where h_0 is the initial thickness of shell ; $\varphi(t)$ is the function describing corrosion wear in time.

The restriction of significant development of vertical cracks was accepted as one of the important criteria, which present a dangerous type of damages, as practice of operation has shown .

Also the environment may favour local conditions for which special care has to be taken in modelling additional flexibilities and non- linearity's. The non- linearity's of cylindrical concrete walls of structures subjected to low -cyclic loadings are mainly the consequences of, partly the opening and reclosure of cracks, and bond slip, material behaviour of reinforcement and for the rest concrete shrinkage and creep [11]. A description of this phenomena is required for: analysed changes in stiffness of the structure; understanding the mechanical behaviour up to the failure. Non-linear material behaviour are modelled by local stress-strain lows. The description of micro cracks (decrease in stiffness) and influence of low- cyclic loadings have been used.

The durability model for service life prediction of concrete silo, bunkers, reservoirs with regard to reinforcement corrosion in conjunction with low cyclic load has also been considered [3]. The influence of environmental conditions is characterized by the external deterioration of concrete and corrosion of steel reinforcement due to carbonation and / or chlorides (Figure 2 a). The durability design has been considered as two- stage hierarchical procedure on the base of numerical modelling. Computation were performed with the help of a program system " LIRA - WINDOWS ", which allows to estimate in axial symmetric and nonsymmetric formulation. The non-line rectangular shell elements (480 elements) have been applied (Figure 2 b).

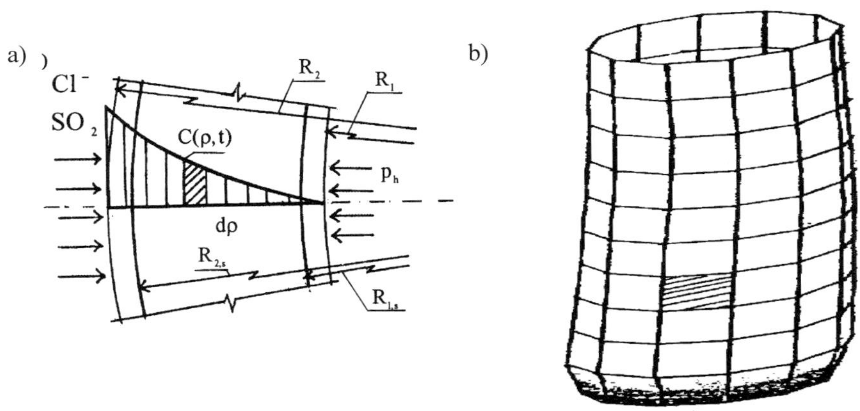

Figure 2 a) model of load and environmental effects on a wall; R_1, R_2 - external and internal radiuses; b) finite element discretisation of cylindrical wall of silo, the damaged element due to corrosion with conjunction low -cyclic load is shaded

The design parameters of the structure are: total heigh of structure H=30 m; radius of circular cross-section r=3.0 m; height of wall section h = 0.2m; concrete compressive strength = 18.9 N/mm²; strength of reinforcement steel = 285 N/mm².

Transformed significances of design parameters were used. The influence of local damages was characterized by increase of ring tensile stresses: 8.2, 8.2, 12.0, 14.8 N/mm² at appropriate terms of operation 0.0, 6.0, 15.5, 34.2 years.

The results of the structure calculation show that short -term and long -term redistribution of inner forces and moments (secondary effects) play an important role and lead to the early local cracking and reduced general crack resistance of structure. It will enable the evaluation of the rate of deterioration due to load and environmental action and to reveal the consequences of failure of cylindrical walls of silos .

REFERENCES

1. FOLIC, R, PAVLOVIC, P, TATOMIROVIC, M, PADONJANIN, V. Repair of soya silo capacity 5400t. Congress FIP 98. Yugoslav National Report Vol 25 No 1 Belgrade 1998 Proceedings ISM Institute. pp 53 - 68

2. POUKHONTO, L M. Studies of concrete silos, bunkers , reservoirs Damage assessment, Modelling and service life prediction. Proceedings of the Radical Concrete Technology.Inter. Congress: Concrete in the Service of Mankind. 24- 28 June 1996. Dundee, Scotland, UK

3. POUKHONTO, L M. Durability Model for Service Life Prediction of Concrete Silo, Bunkers and Reservoirs With Regard to Reinforcement Corrosion in Conjunction With Low Cyclic Load. International Conference on Maintenance & Durability of Concrete Structures , March 4 - 6, 1997, JNT University, Hyderabad, India. pp.235-238

4. SARJA, A and VESICARY, E. Durability design of concrete structures . Report of RILEM Technical Committee 130-csl. Edited by. E & SPON. p.165

5. MEHTA, R K. Concrete technology at the crossroad. Problems and opportunities . ACI, SP - 144,, pp 1 - 30

6. RASTORGUEV, B S, POUKHONTO, L M. Recommendation for the Design of Reinforced Concrete under Action of Short Term and Long Term Low - Cyclic Load : Main Aspects of Standard. Design Methods of Reinforced Concrete Structures. Collection of Works of Moscow State University of Civil Engineering,M.:, pp. 70-84

7. POUKHONTO, L M. Durability of spatial reinforced concrete structures of silos, bunkers and reservoirs . Proceedings of the Inter. Congress IASS - 98 " Spatial Structures in New and Renovation Projects of Buildings and Constructions ": . 22- 26 June 1998. Moscow , Russia .Vol 2,pp. 836 - 843

8. CLIFTON, R J. Predicting the service life of concrete .ACI Materials Journal , November- December 1993, No.6, pp. 611-617.

9. BILCIK, J. Prediction of service life with regard to reinforcement corrosion. Slovak journal of civil engineering ,vol. 3, / 1994 &3 (Bratislava)

10. TANNER, P, ANDRADE, C, RIO, O, MORAN, F. Toward a consistent design for durability. Institute of Construction Science IETcc - CSIC, Madrid, Spain. Proceedings of the Congress FIP 98 . Amsterdam . 1998 pp 1025 - 1028

11. MAZARS, J, CHAVAMIAN, S, RAGUNEAU, F, EFICOS. A Simplified finite element technique in predicting the behaviour of R.C. structures under severe loadings. The French technology of concrete Congress. De La FIP 98, Amsterdam, AFRC- AFREM, 1998, pp 59-72

12. ROGER, S, CROUCH, NENAD, BICANIC. The Use of Non- Linear Finite Element Analysis in Service Life Prediction for Concrete Structures. Proceedings of the International Seminar " Management of Concrete Structures for Long - Term Serviceability", the University of Sheffield . 12 -13 November 1997, pp 57- 73

AN EFFECTIVE PROCEDURE FOR COMBINING ACTIONS

U Quast
Technical University of Hamburg-Harburg
Germany

ABSTRACT. The combinations which may be decisive for the dimensioning of cross-sections can be directly determined by adding the vectors of the action effects in the sequence of decreasing load eccentricity. The simplified combinations which are allowed in building structures are not easy to apply, are not really helpful and should be dropped because they may give more unfavourable as well as more favourable results. A worked example shows the application of the alternative procedure by determining the required reinforcement for a slender two storey column. This column system is transferred to the model column by using a model column coefficient K_M. For computing and plotting the mathematics program EULER is used.

Keywords: Combination of action effects, Reinforced concrete column, Model column, Second order theory, Simplified N/M-interaction, Dimensioning, Mathematics program.

Univ-Prof Dr-Ing Ulrich Quast has been Director in Concrete Structures at the Technical University of Hamburg-Harburg, Germany, since 1985. He was previously a consultant in civil engineering, and became a professor in concrete structures at the Technical University of Braunschweig in 1977. He specialises in theory and design of concrete structures, non-linear analysis, especially for stability endangered slender elements. He serves on different Technical Committees, and he is a past chairman of the stability committee of the former Comité Euro-International du Béton (CEB).

INTRODUCTION

A lot of criticism against the Eurocodes arises from the preconceived idea, that the verification of the general or fundamental combination rule is too complicated. Therefore simplifications of the general rule as given in Eurocode 1 [1], „Basis of Design", by equation (9.10) are deemed to be absolutely necessary. For this reason simplified rules for building structures are given by equations (9.13) and (9.14). These equations are the equations (2.7(a)) and (2.8(a)) and (2.8(b)) in Eurocode 2 [2], "Design of Concrete Structures". In concrete structures longitudinal forces may act favourable or unfavourable, especially with respect to the required amount of reinforcing steel. This fact also complicates the situation as it is.

From Table 1 it can be seen that the total number of possible combinations p really increases very much with the increasing number q of variable actions which are independent from each other. It can also be seen that this number p is significantly reduced by the simplification only in cases with 3 and more variable actions. But the remaining number of possible actions still remains too great. It can be concluded that the reduction of the number of possible combinations is not yet an effective simplification. To determine the decisive combination for cross-section design with 3 variable actions from the totality of 16 simplified combinations is not yet more comfortable than to determine them from 26 combinations. For practical purposes a more pronounced reduction is expected when speaking of a simplification. Alternatively an effective procedure has to be applied, in order to concentrate on the decisive combinations.

For the dimensioning of reinforced cross-sections an effective procedure is to combine the combination of actions with the determination of the required reinforcement. This can be done with computer programs and in the same way by using design charts or other design tools, as for example given in [3].

Table 1 Combination of actions for ultimate limit state design in persistent or transient design situations

COMBINATION OF ACTIONS TO EC 1, Ch. 9.4		q					
		0	1	2	3	4	5
Ch. 9.4.2, eq. (9.10)	$p = 2 + q \cdot 2^q$	2	4	10	26	66	162
$\Sigma\,\gamma_G \cdot G_k + 1.5\,(Q_{k,1} + \Sigma\psi_{0,i}\,Q_{k,i})$, $i > 1$	$r \leq q\,(q+3)/2$	–	2	5	9	14	20
Ch. 9.4.5, simplified verifications for building structures, eq. (9.13) or (9.14)	$p = 2^{q+1}$	(2)	(4)	8	16	32	64
(9.13) $\Sigma\,\gamma_G \cdot G_k + 1.5\,Q_{k,1}$ (9.14) $\Sigma\,\gamma_G \cdot G_k + 1.35\,\Sigma\,Q_{k,i}$, $i > 1$	$r \leq 3\,q$	–	(2)	6	9	12	15

q number of different variable actions, independent from each other
p number of all the possible combinations
r number of the reduced set of combinations which have to be considered for cross section dimensioning by accumulating the action effects in the sequence of decreasing load eccentricity

In both cases the actions are added as vectors within a N/M-chart. The envelope with the most unfavourable and decisive combinations is directly obtained by adding the actions in the sequence of decreasing load eccentricity. This is shown and can easily be understood by giving an example and by explaining the results. For this aim the column shown in Figure 1 with 3 variable actions is analysed.

THE MODEL COLUMN COEFFICIENT K_M

The well known effective column length for buckling design purposes of an isolated element is determined from the equivalence of the buckling load of the real system and of the isolated element. This fundamental idea can also be applied for using the model column method for other columns than real cantilever columns or pin ended columns with the corresponding effective buckling length. The equivalent model columns have to be determined with respect to equal effects of structural deformations.

For the chosen example in Figure 1, consisting of a combination of a pin ended reinforced concrete column and a cantilever steel column, the two different model columns b) and c) in Figure 2 can be derived for the real column system a). In all the three systems the same structural deflection e_2 at the top has to occur so that the same second order effects result for dimensioning the reinforced cross-section b in span 2 of the concrete column.

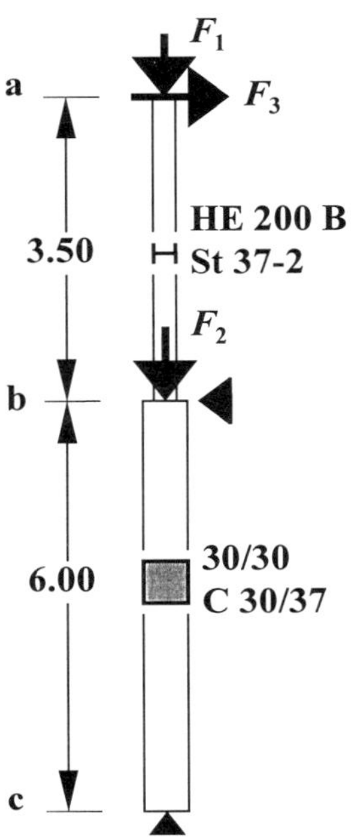

F_i	action	$\psi_{0,i}$	$\gamma_{F,inf}$	$\gamma_{F,sup}$	$F_{k,i}$ [kN]
G_k	permanent				
1	dead load	-	1.0	1.35	-70
2	dead load	-	1.0	1.35	-260
Q_k	variable				
1	snow load	0.6	-	1.5	-30
2	imposed load	0.7	-	1.5	-115
3	wind load	0.6	-	1.5	18

Combination values $\psi_{0,i}$ as given in the German National Application Document.
Partial safety factors $\gamma_{F,inf}$ and $\gamma_{F,sup}$ for favourable and unfavourable effects of the characteristic values of permanent and variable actions G_k and Q_k.

Figure 1 Column data, materials and actions

The expression for the top deflection e_2 can be seen from Figure 2. The model column b) in Figure 2 with the same column length l_1 and the modified curvature K_M $1/r_2$ is taken here for the application of the mathematics program EULER [4].

The results are graphically shown in Figure 4. The model column c) with the same curvature $1/r_2$ and the modified column length $l_1\sqrt{K_M}$ allows to use standard design charts [3].

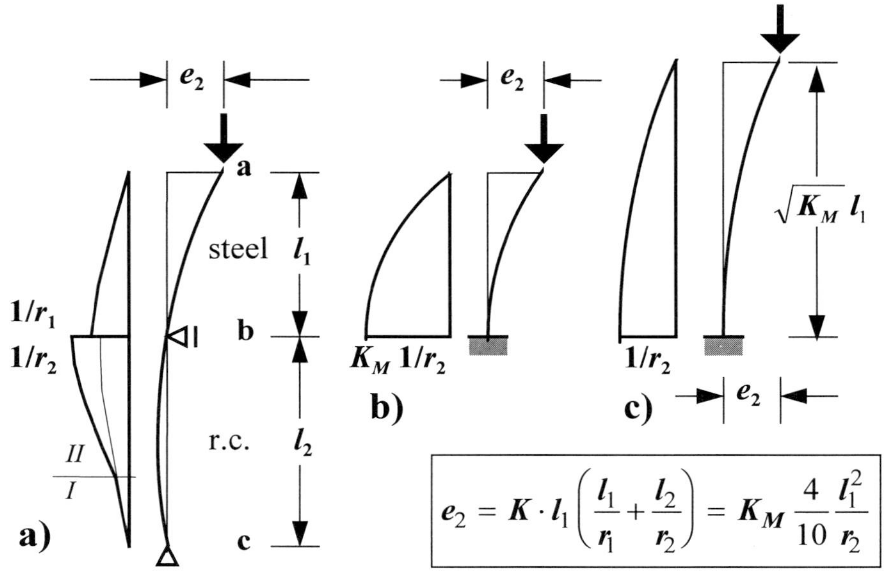

Figure 2 Equivalent model columns for dimensioning the section b in span 2,
a) column as given,
b) model column with the same length l_1 and modified curvature $K_M\ 1/r_2$,
c) model column with the same curvature $1/r_2$ and modified column length $l_1\sqrt{K_M}$.

Instability of statically determined slender columns occurs when yielding in the most stressed cross section happens, which in Figure 2 is section b. The top deflection e_2 can directly be calculated from the curvatures $1/r_1$ and $1/r_2$ at yielding.

For the steel column the influence of the longitudinal force N is unimportant and by not considering it the overestimation of the curvature at yielding is very small.

$$r_1 = (h/2)/\varepsilon_{ay} = 0.5\ h\ E_a / f_{ay} = 0.5 \cdot 0.2 \cdot 210\,000 / 240 = 87.5 \text{ m}.$$

For the reinforced concrete column r_2 can be determined as given by eq. (4.72) in Eurocode 2 with the coefficient $K_2 = f(N_d, A_s) = 1$ because of $|N_d| < N_{bal}$.

$$r_2 = 0.9\ d / (2\ \varepsilon_{yd}) = 0.45\ d / (0.0025/1.15) = 207\ d = 207 \cdot 0.255 = 52.8 \text{ m}.$$

Assuming triangular diagrams for the curvatures, which in this example is on the safe side for the concrete column because of the limited moment magnification, the top deflection is obtained acc. to the corresponding expression in Figure 2 with the coefficient $K = 1/3$,

$$e_2 = (1/3) \cdot 3.50\ (3.50/87.5 + 6.00/52.8) = 0.179 \text{ m} = (4/10)\ K_M\ 3.50^2/52.8,$$

which then yields the model column coefficient K_M for this example as $K_M = 1.93$.

THE COMBINATION OF ACTIONS AND DIMENSIONING

Simplification of the N/M-Interaction

In addition to computer programs and design charts simplified methods may nevertheless be useful, be it to facilitate better understanding of fundamentals or be it to have an alternative method for checking computer results. Not to forget that the usage of design charts sometimes is uncomfortable: Special parameters have to be determined, interpolation between different charts may be necessary, due to the limited size of charts the accuracy for taking values is sometimes limited too, and so on.

The described simplification is valid for symmetrically reinforced rectangle cross sections. In the ultimate limit state both reinforcements in a wide range are plastically stressed and therefore the sum of their forces yields zero as expressed by eq.(1) in Figure 3. This may be assumed for $|N| \leq N_{bal} = N_{0c}/2$ and $h_1/h \leq 0.15$. The resultant of the concrete stresses N_c has to be equal to the acting longitudinal force N. From eq. (2) this allows to determine the depth x_B of the compression stress block acc. to eq. (4) and the bending moment M_c of the concrete stresses with eq. (3) as a function of N. The formulae can finally be derived such that the required reinforcement can directly be computed:

$$req\, A_s\,(N, M) = \frac{M - M_c(N)}{z_{s1}\, f_{yd}} = \frac{M - N\dfrac{h}{2}\left(\dfrac{N}{-b\,h\,0.85\, f_{cd}} - 1\right)}{z_{s1}\, f_{yd}}$$

In comparison with exact solutions the differences are about ±3 % only.

Adding the Vectors in the Sequence of Decreasing Load Eccentricity

For the combination of actions together with the simplified dimensioning of reinforced cross sections the mathematics program EULER [4] is used for doing the calculations and also to generate the plot of Figure 4. The formulae with their results can be seen in the Appendix [5].

The second order analysis of a column is transformed to cross section design acc. to ch. 4.3.5.6.3 (b) of Eurocode 2. The total design moment $M_{Sd,tot}$ is the sum of the first order moment $M_{Sd,0}$ augmented by $M_{Sd,a}$ allowing for the effect of imperfections and by $M_{Sd,2}$ allowing for the effect of structural deformations, the second order effect as it is called,

$$M_{Sd,tot} \qquad = M_{Sd,0} + M_{Sd,a} + K_2(N_{Sd}, A_s)\, M_{Sd,2}$$

The characteristic values of $M_{Sk,0}$, $M_{Sk,a}$ and $M_{Sk,2}$ for $K_2(N_{Sd}, A_s) = 1$ can be seen in the Appendix [5]. With respect to the cross section b there are no effects of imperfections and of structural deformations within the action effects from G_2 , Q_2 and Q_3 .

The results are graphically shown within a detail of the N_d/M_d-chart in Figure 4, which shows the design values $\gamma_G\, G_k$ and $\gamma_Q\, Q_{k,i}$ of the action effects as vectors. On the corresponding vectors of the variable action effects also the design values of the combination values

$\gamma_Q\ \psi_{0,i}\ Q_{k,i}$ are marked by diamonds. The starting point of all the vectors of the variable action effects is the value 1.00 G_k of the permanent action. The part 0.35 G_k, which has to be added in cases when being unfavourable, appears like a variable action.

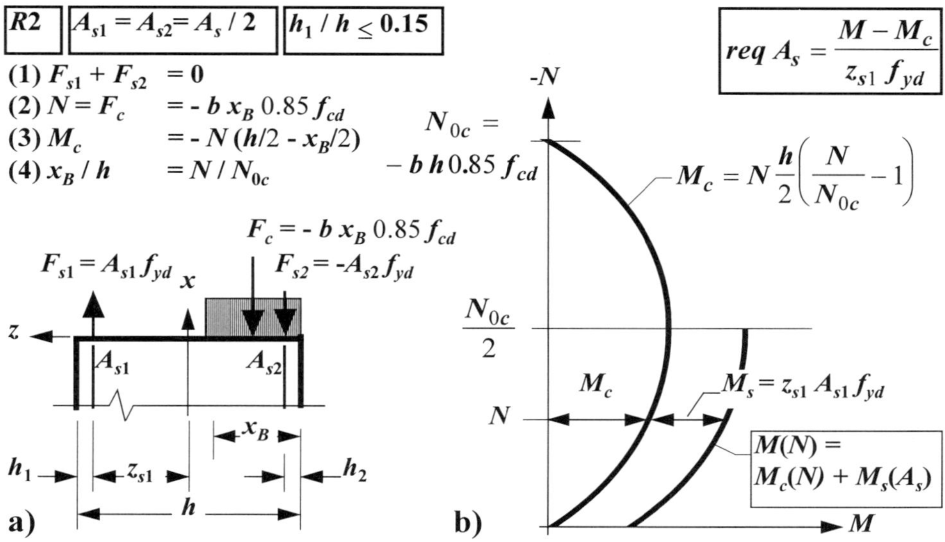

Figure 3 Simplified N/M-interaction for rectangle cross sections

The decisive combination in Figure 4, giving the maximum required reinforcement area, is a point on the boundary which is formed by the polygone

$$1.0\ G_k + 1.5\ Q_{k,3} + 1.5\ \psi_{0,1}\ Q_{k,1} + 0.35\ G_k + 1.5\ \psi_{0,2}\ Q_{k,2}$$

This boundary is formed by the permanent action effect, the dominant variable action effect, the other variable action effects and the 0.35fold permanent action effect in the sequence of decreasing load eccentricity $e_d = M_d/|N_d|$.

Which point on this boundary gives the greatest required reinforcement depends from the greater or smaller inclination of the N_d/M_d-line, as it is the case for different arrangements of the reinforcement in the cross section, for example at four sides instead of only two sides as in this example.

Especially this point needs not be the point with the greatest axial force max |N|, nor the point with the greatest bending moment max M, nor the point with the greatest load eccentricity max e, as can be seen from Figure 4.

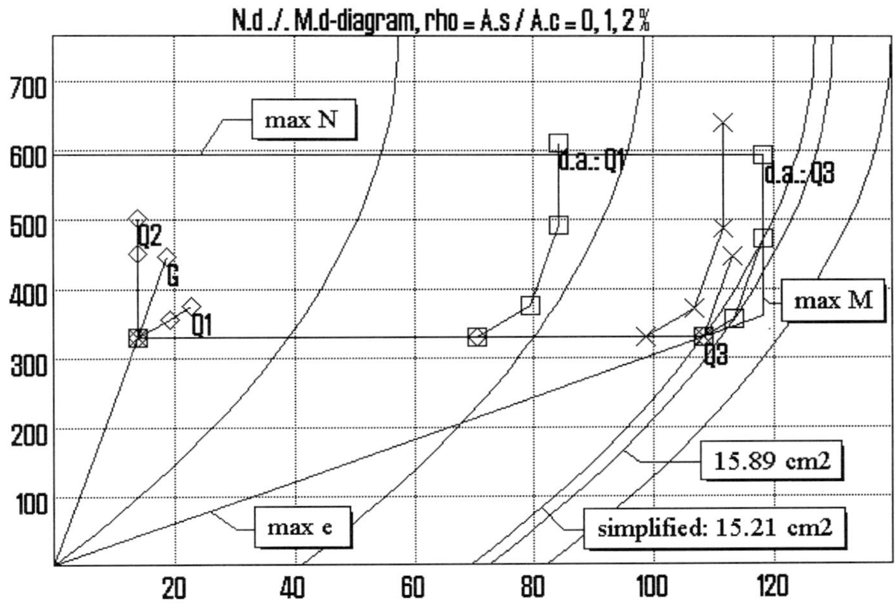

Figure 4 Action effects and cross section capacity

Results

In this example Q_3 is not the dominant action because it yields the most unfavourable action effect, as it can clearly be seen from Figure 4. Q_3 is in this example the dominant action because its reduction $\gamma_Q (1 - \psi_{0,i}) Q_{k,i}$ if not being the dominant action is the most unfavourable one compared with the possible reductions of the other variable actions Q_1 and Q_2. These possible reductions of the variable action effects are the distances between the marked points and the end of the vectors of these variable action effects. When having a graphical representation of the action effects the dominant action i can in many cases be detected as that action which has the most unfavourable part $\gamma_Q (1 - \psi_{0,i}) Q_{k,i}$. Only for demonstration purposes and to make comparisons possible the boundary belonging to Q_1 assumed to be the dominant action is also shown in Figure 4.

In this example longitudinal forces N act favourably. The decisive fundamental combination is therefore $1.00\ G_k + 1.5\ Q_{k,3} + 1.5 \cdot 0.7\ Q_{k,1}$, which is load case 3 within the results to be seen in the Appendix [5].

The decisive simplified combination is $1.00\ G_k + 1.5\ Q_{k,3}$. It is more unfavourable than the simplified combination $1.00\ G_k + 1.35\ Q_{k,3} + 1.35\ Q_{k,1}$. It requires 15.21 cm^2 reinforcing steel, which are only 96 % of the reinforcing steel of 15.89 cm^2, which is required for the fundamental combination. The simplification is not on the safe side, even if the difference is not significant from the practical standpoint. The difference of the simplified design procedure from Figure 3 referred to the exact solution of 16.29 cm^2 is only 2.5%.

The number r of the reduced set of combinations which form the possible boundaries and which are in general sufficient to be considered and which have to be considered only then, if the dominant variable action is not known in before, are given in Table 1. There is nearly no difference between the fundamental combinations and the simplified ones. Knowing in this example that Q_3 is the dominant variable action, the corresponding boundary in Figure 4 needs to consider up to 5 load cases only from the totality of 26, whereas there are two corresponding boundaries for the simplified combinations with up to 6 load cases from the totality of 16. It can be concluded that the simplified verification of the combination of actions for building structures acc. to eq. (9.13) and (9.14) in Eurocode 1, which are the eq. (2.8(a)) and (2.8(b)) in Eurocode 2, is not really simpler. This so-called simplification should therefore be taken away. The advantage would be, that equivocal and contradictory designs are avoided and that it becomes very obvious, that a comprehensible procedure has to be applied.

CONCLUSIONS

1. The most unfavourable portion $(1 - \psi_0)\ Q_d$ of variable actions determines the dominant one.
2. Accumulating the action effects in the sequence of decreasing load eccentricity e yields the decisive load cases for dimensioning.
3. The simplified combinations are ineffectual because:
   - the number of load cases to be treated does not decrease.
   - to reduce only the totality of possible combinations is not sufficient, here from 26 to 16.
   - the results may be on the safe as well as on the unsafe side.
4. The load cases with greatest longitudinal force max $|N|$, greatest bending moment max M or greatest load eccentricity max e need not be the decisive ones.
5. The decisive combination also depends on the inclination of the N/M-interaction, according to the arrangement of the reinforcement.
6. The content of tables can be better visualised by graphs, leading to better explanation and understanding of the fundamentals and inter-relationships.
7. Simplified formulae for the N/M-interaction allow the direct computation of the required reinforcement and render superfluous the usage of charts with the well known inconveniences.

REFERENCES

1. ENV 1991-1. Eurocode 1: Basis of Design and Actions on Structures, Part 1: Basis of Design. CEN, 1993.

2. ENV 1992-1-1. Eurocode 2: Design of Concrete Structures, Part 1-1: General Rules and Rule for Buildings. CEN, 1991.

3. QUAST, U. Dimensioning of Slender Elements Related to Ultimate Limit States Influenced by Structural Deformations - Stability Check. Concrete Structures - Euro-Design Handbook 1994/96, Ernst & Sohn, Berlin, 1995, pp 453-513.

4. GROTHMANN R. Euler for Windows 95, freeware, 1998. hhtp://mathsrv.ku-eichstaett.de/MGF/grothmann/euler.html,

5. QUAST, U. Appendix with the complete listing of the computation for the column example in Figure 1. hhtp://www.tu-harburg.de/mb/DundeeQsAppendix.html.

APPENDIX

The computations for the column in Figure 1 are done with EULER for Windows 95, © R Grothmann [4]. EULER is a program for quickly and interactively computing with real and complex numbers and matrices, or with intervals. The computation can be downloaded [5].

RESEARCH TO SUPPORT THE DEVELOPMENT OF EUROPEAN STANDARDS FOR TESTING CONCRETE

D C Spooner

British Cement Association

UK

ABSTRACT. The paper is concerned with the resolution of some technical issues arising during the drafting of a series of European standards for concrete testing. The issues which needed attention were due to differences in national and international procedures and practices, and which, because of the absence of relevant technical data, were not able to be resolved by discussion in committee. The technical debate concerned the manner of production of specimens for testing and the apparatus and procedures for determining certain properties of both fresh and hardened concrete. It was decided that an inter- laboratory comparison was required. The comparison involved nine European laboratories which specialised in concrete testing; some national laboratories, others commercial laboratories.

As a result of the findings of the experimental work and the comments received during CEN enquiry, the draft standards were revised. They have now been sent for formal vote, prior to becoming European standards.

Keywords: Concrete, Properties, Testing, Comparison, Laboratories, Standardisation.

Professor David C Spooner is Director of the Materials and Standards Division of the British Cement Association, has researched many aspects of concrete testing and performance and is convenor of CEN/TC 104/TG 8. He is a visiting Professor at the University of Sheffield.

INTRODUCTION

When the European pre-standard for concrete, ENV 206, was published in 1990 it called up the then ISO standards for methods of testing concrete. Subsequently, the technical committee responsible for drafting the full standard, EN 206, took the decision to revise these ISO testing standards and to set up a task group, CEN/TC 104/SC 1/TG 8, Test Methods, to undertake the work. This task group met for the first time in 1991, in the UK, and commenced discussion on the revisions of, in total, eighteen standards.

The process of CEN standardisation firstly involves the production of a draft for 'enquiry' which establishes whether, in principle, EU members are willing to accept the draft. This is followed by a second stage of formal vote for approval, or not. Once approved, all EU members are obliged to implement the Standards in their country. If the Standard is a full standard (an EN) then conflicting national Standards must be withdrawn. If the standard is an ENV then the standard is voluntary and conflicting national Standards are not required to be withdrawn.

Task group TG 8 was charged by the Technical Committee SC 1 with producing a suite of ENs on test methods for fresh and harden concrete. Thus the group firstly set about drafting text for 'enquiry' drafts. Discussions on each of the test methods were based on the original ISO text, modified on the basis of the expertise of each of the group members. Experts who were appointed to TG8 came from the following European Countries:

Denmark (DK)	France (FR)
Finland (F)	Germany (DE)
Greece (GR)	Italy (IT)
Netherlands (NL)	Portugal (PT)
Sweden (SE)	United Kingdom (UK)

It became clear to TG 8 during discussions on the content of the Standards that there were differences between the national Standards and between those and their ISO equivalents, that were, in part, due to custom and tradition rather than to technical reasons. That being so it was impossible, in some instances, to reach a decision based on technical arguments. It was therefore decided, for the enquiry drafts, to allow a number of options if no technical preference was obvious and, in the mean time, to try to undertake some laboratory work to compare these alternative methods of test.

An application was made in 1994 to the Measurement and Testing programme of DGX11 of the EC for funding to part support the work. The proposal was successful and the balance (50%) of the funding was provided by the partners in the project.

The project was entitled 'Inter-laboratory comparisons in Support of CEN-ISO Standard called up in EN 206, Concrete'. It commenced in December 1994 and produced the final report in June 1997. The partners in the project were, in the main, members of the Task Group TG 8, or were nationally recognised laboratories in the field of concrete testing.

PARTNERS IN THE PROJECT REF: MATI-CT94-0043

- British Cement Association (BCA), (UK) lead partner
- Centre Experimental de Recharches et d'Etude du Batinent et Travaux Publics (CEBTP), (FR)
- Central Laboratory for Public works (KEDE), (GR)
- DTI Building Technology (DK)
- Laboratori Nacional de Enganharia (LNEC), (PT)
- Eur. Ing R Anderson (UK)
- Mr M Roberts, C&G Concrete Ltd, (UK)
- Building Research Establishment (BRE Ltd), (UK)
- Centre for Civil Engineering, Research and Codes (CUR), (NL)
- Centro Technico Gruppo SpA (CTG), (IT)
- Green Land Reclamation Ltd (UK)
- Prüftechnik IFEP GmbH, (DE)
- Swedish National Testing and Research Institute (SP), (SE)

OBJECTIVES OF THE LABORATORY WORK AND EXPERIMENTAL APPROACH

In its deliberations, TG8 identified a number of procedural differences between European countries which had arisen because of national traditions and preferences. TG8 was unable to chose between these on an objective technical basis and proposed that an experimental programme be undertaken to shed light on the problems.

The procedural differences identified by TG8 were:

(1) *The method of curing concrete after stripping the concrete from the mould:*

- in water
- in water then in air
- in a high-humidity chamber.

(2) *The method of compaction of fresh, plastic, concrete into moulds to produce specimens:*

- using a vibrating table on which moulds are placed
- using a poker-shaped vibrator (internal vibrator) inserted into the fresh concrete
- compaction by hand using a light steel rod
- compaction by hand using a heavy steel bar.

(3) *The method of treating the ends of concrete cylinders to produce flat surfaces, particularly for strength measurement:*

- grinding flat
- capping with sulfur mixture
- capping with high-strength sulfur mixture
- capping with high-alumina cement mortar
- capping with special end caps containing sand (sand-box).

(4) *In the measurement of tensile splitting strength, the means of applying the force*:

- flat or curved loading pieces
- hardboard or steel packing strips.

(5) *The load application in the measurement of the flexural strength of concrete*:

- two lower supports to the beam and a central top loading (centre-point loading)
- two lower supports to the beam and two top loading points (two-point loading)

(6) *The type of air-meter used to determine the entrained air content in fresh concrete:*

- water column meter (Type A)
- pressure gauge meter (Type B)

(7) *The influence of the core diameter on the measured compressive strength when cores are drilled from hardened concrete structures:*

- effect of core diameters of 25,50,100mm
- influence of aggregate size in the concrete

The experimental programme was devised to provide guidance on the questions raised in the previous section, as follows:-

Experiment A: the investigation of the effects of the method of curing, and shape and size of test specimens on the compressive strength and hardened density of concrete test specimens.

Experiment B: the investigation of the effects of the method of compaction, and shape and size of test specimens on the compressive strength and hardened density of concrete specimens.

Experiment C: the investigation of the effects of different methods of capping and methods of curing, on the compressive strength of concrete specimens.

Experiment D: the investigation of the effects of the type of loading piece, the materials of packing strips and shape and size of test specimens on the tensile strength of concrete specimens.

Experiment E: the investigation of the effect of centre-point or two-point loading and prism size on the flexural strength of concrete specimens.

Experiment F: the investigation of the effect of the type of air meter on the measurement of the air content in concrete specimens.

Experiment G: the investigation of the effects of different core diameters on the compressive strength of concrete specimens.

The experiments were carried out in nine participating laboratories, each in a different European country. The laboratories were given detailed instructions, which were discussed at the first participants meeting, regarding procedures to follow to ensure repeatability conditions. Consideration was given to the possibility of standardising materials throughout the programme but dismissed as being impracticable and costly. Laboratories were therefore provided with performance specifications which had to be met for the concretes used in each experiment. The choice and supply of materials and the mix designs, were left to individual laboratories to decide. The purpose in requiring the laboratories to produce the concrete mixes to agreed performance specifications was to facilitate later statistical analysis into separate, practical, groupings of results, such as, for example, strength level (150 mm cubes), slump, air content, aggregate size or combinations of these. Each experiment comprised a 'Trial 1', 'Trial 2' and 'Test'. The object of the trials being to familiarise laboratories with procedures and methods and to ensure that performance targets were met. In most cases it was found that the quality of the data reported for the trials was high enough to be included in the data on which the conclusions are based.

The assignment of experiments to laboratories was as follows:-

Experiment/Laboratory	A	B	C	D	E	F	G
AC1-GB	√			√		√	
AC2-FR	√		√				√
AC3-DE	√			√			√
AC4-PT	√			√		√	
AC5-SE		√	√				
AC6-IT		√			√	√	
AC7-DK			√		√		
AC8-NL		√			√		√
AC9-GR	√	√					

RESULTS AND DISCUSSION

The Method of Curing Concrete After Stripping From the Mould

The effect of the method of curing was examined on specimens which were all demoulded at day 1 after casting. They were then (a) immersed in water at 25°C ± 2°C or (b) immersed in water until day 8 and then placed in air at 15-22°C and 60-70% RH or (c)placed in a humidity chamber at 20°C ± 5°C and >95% RH. All the specimens were compacted using 25 mm diameter poker vibrators. The ends of cylinders were ground flat at day 25 after which the specimens were returned to the respective curing condition. The specimens were tested for compressive strength at day 28.

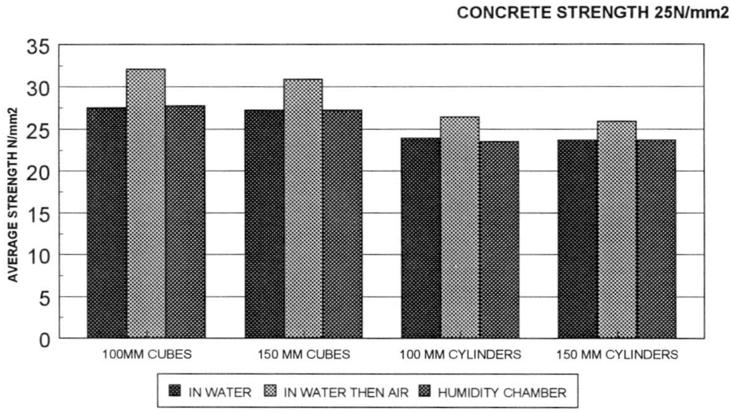

Figure 1 Effect of method of curing on strength for cubes and cylinders 25 N/mm^2 concrete

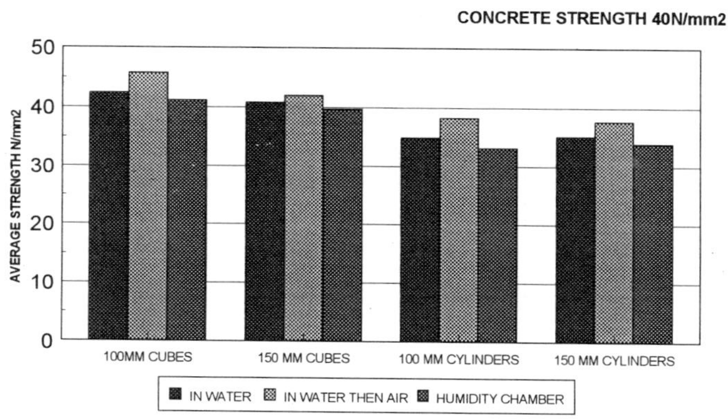

Figure 2 Effect of method of curing on strength for cubes and cylinders, 40 N/mm^2 concrete

Five laboratories participated in this experiment. Each laboratory made a total of 24 specimens from each of the 'Trial 1' 'Trial 2' and 'Test' batches, 6 specimens each of 100 mm cubes, 150 mm cubes, 100 mm diameter x 200 mm cylinders and 150 mm diameter x 300 mm cylinders. This was repeated for concrete of nominal strength 25 N/mm^2 and 40 N/mm^2. Therefore each of the average results for the strength levels shown in Figures 1 and 2 are for 30 specimens.

The results show that for both levels of concrete strength curing in water then in air results in significantly higher strengths by some 10%, for both cubes and cylinders. Curing in water or in a high humidity cabinet yields very similar results.

For specimens cured in water or in a high humidity cabinet the results show that 100 mm diameter cylinders give strengths which are about 84% of 100 mm cubes and 150 mm diameter cylinders give strengths of 87% of the 150 mm cubes. The data also suggests that 100 mm cubes are some 3% stronger than 150 mm cubes but that there is very little difference between the strengths of the two sizes of cylinders.

The Method of Compaction of the Fresh Concrete

The methods of compaction which were examined for their effect on strength were internal vibrator (25 mm diameter poker), vibrating table, compacting rod and compacting bar. Two of both cubes and cylinders were used, two levels of concrete strength at one of two levels of slump and four laboratories participated. For each of the three batches (Trail 1, Trial 2, Test) at each level of strength or slump, 32 specimens were made, 8 each of the four mould sizes as used in Experiment A. Of the group of 8 specimens made using a particular mould size, 2 were compacted using each of the four compaction methods. The ends of cylinders were ground flat. Specimens were tested at 28 days.

The results in Figures 3 and 4 for, respectively, the 25 N/mm^2 and 40 N/mm^2 concrete strengths are, in each case, the overall averages from three laboratories at both levels of slump. Thus each mean strength is of 24 specimens.

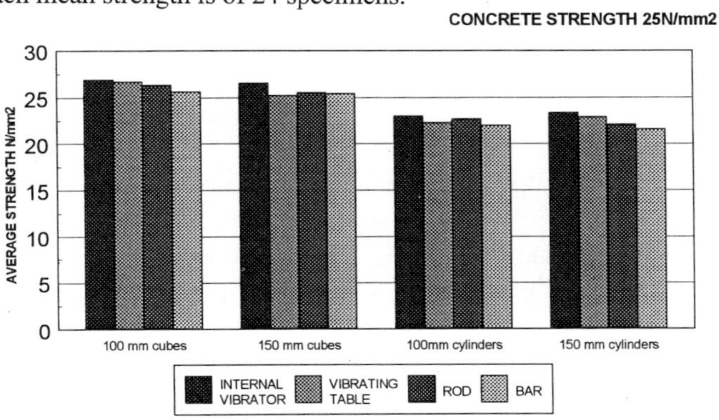

Figure 3 Effect of method of compaction on strength for cubes and cylinders, 25 N/mm^2 concrete

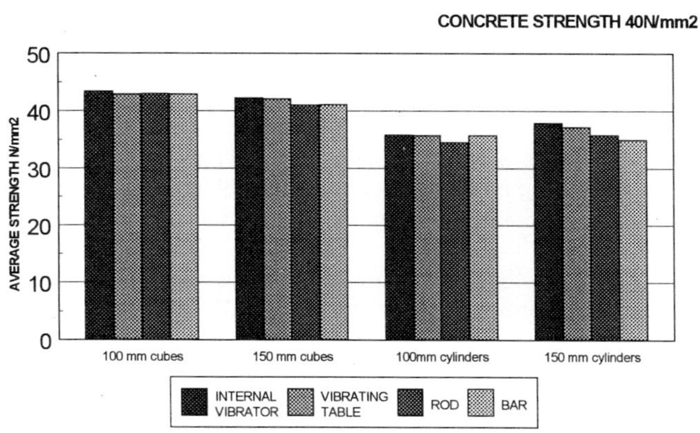

Figure 4 Effect of method of compaction on strength for cubes and cylinders, 40 N/mm^2 concrete

The results in Figures 3 and 4 show that although there may be a tendency for the internal vibrator to result in higher strengths, particularly with the larger cylinders, the effect is relatively small. In general it can be said that all the methods give satisfactory compaction with these two, non-air entrained, concretes. The reason that the averages in Figures 3 and 4 do not include the results from the fourth laboratory is that here, air entrained concrete was tested and the results were significantly affected by the method of compaction. This is illustrated for the 40 N/mm^2 strength concrete in Figure 5.

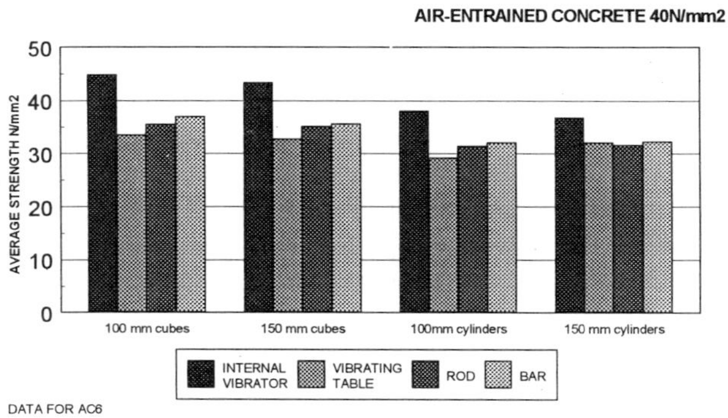

Figure 5 Effect of method of compaction on cubes and cylinders, air entrained concrete

It is evident that the use of the internal vibrator has expelled the entrained air and hence increased the strength. This was confirmed by density measurements.

The Method of Capping or Adjusting the Ends of Cylinders

Three laboratories participated in this experiment, each testing concrete at one of three strength levels, 25 N/mm^2, 40 N/mm^2 or 100 N/mm^2. The specimens were 150 mm diameter x 300 cylinders, cured using the three alternatives described in Experiment A and compacted using an internal vibrator.

A total of 30 cylinders was made from each of the Trial 1, Trial 2 and Test batches, 6 cylinders of each batch being adjusted or capped by grinding, high-strength sulfur capping, high-alumina mortar capping, use of sand-box end caps or ordinary sulfur capping.

The average strengths in the figures which follow are therefore the means of 18 specimens. Figure 6 shows the results from laboratories coded AC2 and AC5 and clearly there is no noticeable effect of the method of capping.

However, laboratory coded AC7 had to test very high strength concrete and had consequent problems with lack of adhesion of some of the capping materials to the concrete. The result was a wide range of cylinder strengths as is illustrated in Figure 7 for two batches of concrete.

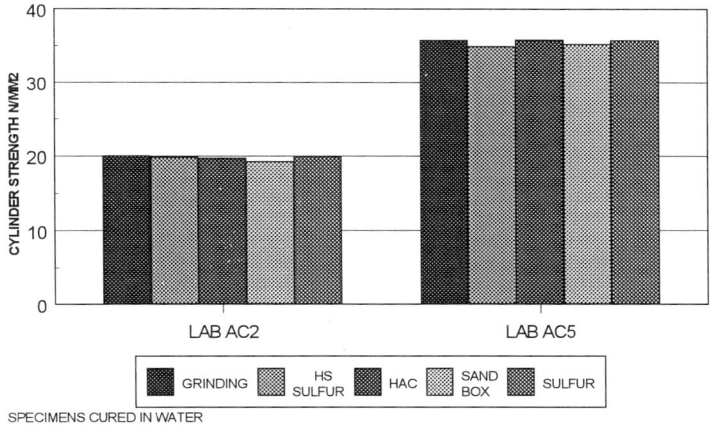

Figure 6 Effect on the strength of cylinders adjusted or capped using different methods

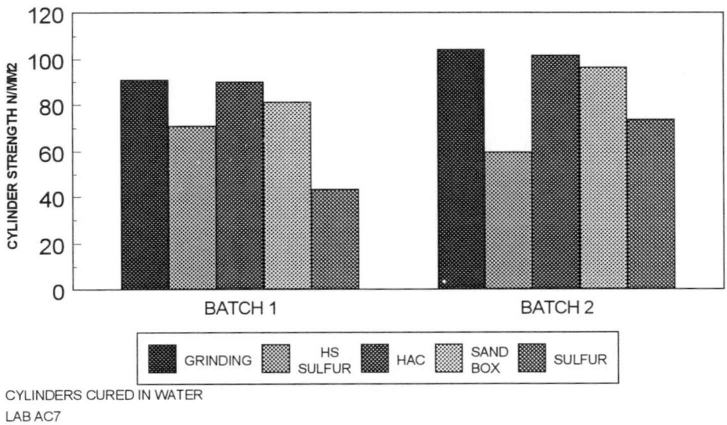

Figure 7 Effect on the strength of cylinders adjusted or capped using different methods (NB. High-strength concrete)

The Shape of the oading piece and the type of packing material on the tensile splitting strength of cylinders and cubes

Three laboratories participated in this experiment, each testing at one of three concrete strength levels, 25 N/mm^2, 40 N/mm^2 and 60 N/mm^2. Each laboratory made 24 specimens from each batch (Trial 1, Trial 2, Test), 6 of each 100 mm and 150 mm cubes and 100 mm and 150 mm diameter cylinders.

All the specimens were compacted using an internal vibrator and were cured in water. The splitting strength was determined using three configurations:

FH = flat loading piece (i.e. the testing machine platen) with hardboard packing strips

CH = curved loading pieces (i.e. specially made curved loading pieces, with 75 mm radius, which provides load transfer to the test specimen) with hardboard packing strips

FS = flat loading piece with steel packing strips

Of the 6 specimens of a particular size and shape made from each batch, 2 were tested in each of the three configurations above. There was no evidence that the level of concrete strength influenced the conclusions and so the tensile strength averages in Figure 8 are for all three strengths. Each mean is therefore of 18 specimens.

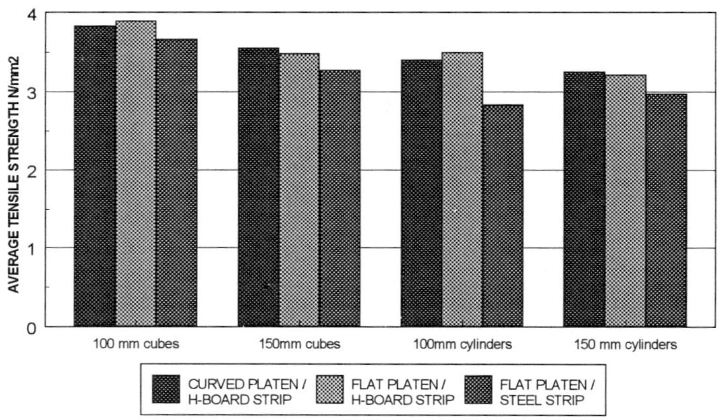

Figure 8 Effect of specimen and load application the apparent tensile strength of concrete

It is evident from the figure that cubes produce higher apparent tensile strengths than cylinders of the equivalent diameter. Also that the use of steel strips produced the lowest results. The ratios of apparent tensile strengths of cubes to cylinders, of equivalent sizes and tested on hardboard, is 1.12 for 100 mm cubes to 100 mm diameter cylinders and 1.09 for 150 mm cubes to 150 mm diameter cylinders. The use of 100 mm cubes with hardboard strips gives results higher by 10% than using 150 mm cubes and for cylinders the corresponding figure with hardboard strips is 7%. The overall mean strengths, using hardboard strips, of loading through flat, as opposed to curved, loading pieces were identical. Thus there seems no advantage (or disadvantage) in the use of the special curved loading pieces.

The Load Application in the Measurement of the Flexural Strength of Concrete

Three laboratories participated in this experiment, testing at one of three concrete strength levels, 25 N/mm^2, 40 N/mm^2 and 60 N/mm^2 and casting both 100 x 100 x 400 prisms and

150 x 150 x 600 mm prisms. The specimens were compacted using internal vibrator and cured in water and 8 of each prism size were made from each of the three batches at each laboratory. Four specimens of each size were tested on either a two-point flexural loading system (two supporting rollers and load applied by two top rollers) or a centre-point loading system (two supporting rollers and load applied by a central, single, roller). There was no evidence of an effect of level of strength on the conclusions and so the mean strengths shown in Figure 9 are for all three strength levels, that is for 36 specimens .

It is evident that centre-point loading gives higher apparent flexural strengths than does 2-point loading. The ratio is close to 1.10 for both specimen sizes. There is also a small but significant difference in apparent strength due to specimen size, with the smaller prisms being stronger by 2.5%. Density measurements, not reported here, show a corresponding higher value by some 0.5%.

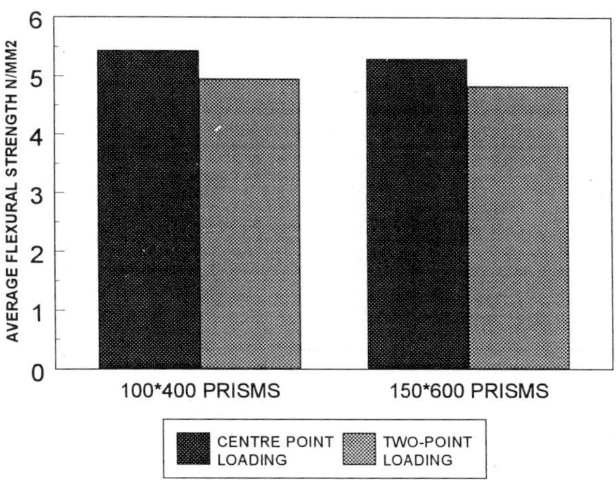

Figure 9 Effect of prism size and loading system on the apparent flexural strength

The Type of Air Meter Used to Measure the Air Content of Fresh Concrete

Three laboratories participated in this experiment each measuring the air contents of three concretes with nominal values of 0%, 4% and 7%. Each laboratory was required to produce concrete of nominal 40 N/mm^2 strength and with two levels of slump, 20 mm and 100 mm. From each batch of concrete (3 air contents, 2 slumps), two separate tests were made using each air meter. This was repeated for Trial 1, Trial 2 and Test, as in other experiments.

The air meters used were either of the water-column type (type A) or the pressure-gauge type (type B). Internal vibrators were used to compact the concrete. The air contents in the data presented in Figure 10 are averages including both slump levels since slump did not significantly affect the results. Each average is thus obtained from 36 measurements. The instruction to laboratories when using mechanical methods of compaction was to "continue compaction until no further appearance of large air bubbles on the surface of the concrete and the surface becomes relatively smooth with a glazed appearance, without excessive segregation."

Two laboratories followed this instruction whilst the third used a constant time of vibration of the internal vibrator. The two laboratories found that the water column meter gave slightly higher results (0.1 to 1.0% air) than the pressure gauge meter.

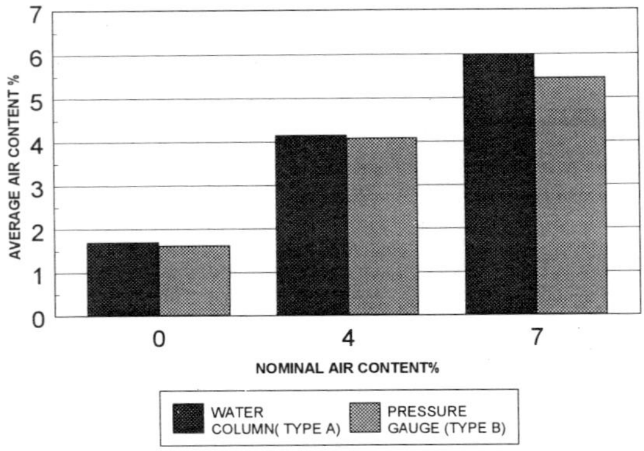

Figure 10 The effect of the type of air meter used for the air content of fresh concrete

The third laboratory found the reverse (less by 0% to 0.3%). In all laboratories the differences were higher at higher air contents. Examination of the recorded times for which vibration was applied , in the two laboratories which found that the water-column meter gave the higher results, showed that the times were shorter for the concrete in the water-column meter than in the other meter. Also there was a correlation between the differences in air content measured by the two meters and the differences in time of applied vibration. A difference in vibration time of 10 secs corresponded to some 0.4-1% in air content. The implication seems to be that the operator's assessment of the achievement of full compaction depended on the particular type of air meter (possibly related to the shape of the bowls?). The use of internal vibrators in compacting air entrained concrete may provide a too severe action and other methods are preferable.

The Influence of Core Diameter on the Compressive Strength of Concrete Cores

Three laboratories participated in this experiment, each casting two 1000 x 250 x 250 mm concrete beams at one of three strength levels, 25 N/mm^2, 40 N/mm^2 or 60 N/mm^2. At each strength level, the concretes were made using either 20 mm or 40 mm nominal maximum size aggregate. The target slump was 100 mm for all mixes. From each of the two beams cast at each laboratory three cores were cut of diameter 25, 50, and 100mm.The cores were cured in water, the ends were prepared by grinding and they were tested at an age of 28 days. This procedure was repeated for Trial 1, Trial 2 and Test. Since the level of concrete strength did not affect the conclusions, the mean compressive strengths shown in Figure 11 are therefore of 27 core results.

It is apparent from Figure 11 that the effect of the diameter of the core on the measured compressive strength is significant. The smaller the diameter the lower the strength. The effect of core diameter on strength is more marked for larger maximum size of aggregate, to the extent that it could be argued that 100 mm diameter cores are adequate for 20 mm maximum size aggregate. However, it is possible that 100 mm diameter cores are still too small for concrete with 40 mm maximum size aggregate. It is also clear is that if cores of 25 mm or 50 mm diameter are used for strength measurements of in- situ concrete, care must be taken in interpretation of the results.

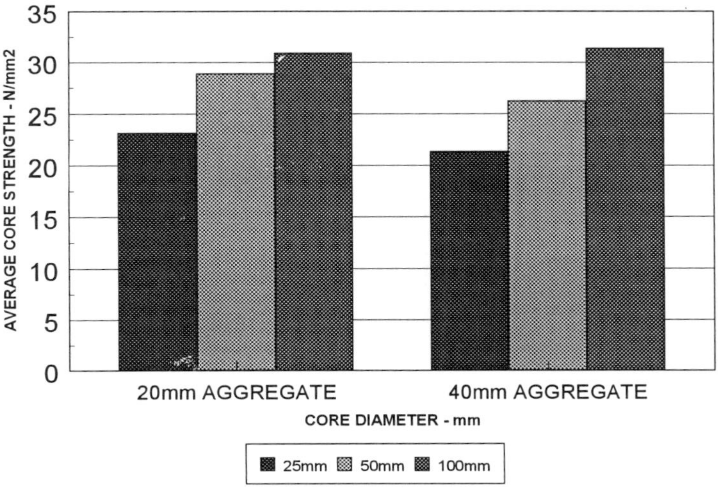

Figure11 Effect of core diameter on measured compressive strength

FINAL REMARKS

It is believed that the experimental work described earlier in this paper has provided important information to assist in the production of more rigorous test methods for concrete.The simple fact of pan-European co-operation in the programme has resulted in a broader acceptance of the results than would have occurred if only one country had done the work. The results of the experiments have been provided to CEN/TC 104/TG 8 and, together with national comments received on the ‘enquiry’ drafts of the testing standards, have resulted in revised versions of the concrete testing standards being sent, in 1999, for formal vote by E.U. Member States.

REVISION OF THE SERVICEABILITY PROVISIONS OF THE AUSTRALIAN STANDARD FOR CONCRETE STRUCTURES – A NEW APPROACH FOR DEFLECTION CALCULATION AND CONTROL

R I Gilbert

The University of New South Wales

Australia

ABSTRACT. In the design of reinforced concrete structures for serviceability, the simplified procedures contained in AS3600-1994 for calculating the deflection of beams and slabs are less than adequate. This is particularly so for deflection sensitive elements such as floor slabs. The calculated deflection is often significantly less than the actual deflection and serviceability problems resulting from excessive deflection are not uncommon for structures designed in accordance with the Standard. The same criticism can be levelled at many design codes. A review of the deflection calculation procedures in AS3600 is currently being undertaken by Standards Australia. In this paper, the deficiencies of the existing approach are discussed and suggestions for improving the procedure are proposed and evaluated. The corresponding provisions in Eurocode 2 are also evaluated.

Keywords: Australian standard, Codes, Creep, Cracking, Deflection, Serviceability limit states, Slabs, Shrinkage, Tension stiffening, Time-dependent.

Professor R Ian Gilbert is Head of the School of Civil and Environmental Engineering at the University of New South Wales. Prior to joining the University, he worked as a structural design engineer for two Sydney based consultants. For the past twenty-four years, he has taught structural engineering and undertaken research in the area of concrete structures. His publications include four books and over one hundred refereed papers in the area of reinforced and prestressed concrete structures. He has served on Standards Australia's Concrete Structures Code committee BD/2 since 1981 and also on the Composite Code Committee BD/32 and he was actively involved in the development of AS3600.

INTRODUCTION

The deflection and extent of cracking in reinforced concrete flexural members under sustained service loads depend primarily on the non-linear and in-elastic properties of concrete and, as such, are difficult to predict with confidence. The problem is particularly difficult in the case of slabs, which are typically thin in relation to the their spans and are therefore deflection sensitive. It is stiffness rather than strength that usually controls the design of slabs, particularly in the cases of flat slabs or flat plates.

In most concrete codes (including the Australian Standard AS3600 [1] and Eurocode 2 [2]), two basic approaches for deflection control are specified. The first is deflection control by the calculation of deflection. The second is deflection control by the satisfaction of a maximum span to depth ratio or a minimum thickness requirement. Code procedures for the calculation of the final deflection of beams and slabs are necessarily design-oriented and simple to use, involving crude approximations of the complex effects of cracking, tension stiffening, concrete creep and shrinkage and the load history.

For beams containing more than about 1% tensile steel ($A_{st}/bd > 0.01$), the simplified deflection calculation procedures in AS3600 are usually adequate. For more lightly reinforced beams and slabs, the procedures are inadequate. The effects of cracking are not adequately modelled, particularly the loss of stiffness resulting from time-dependent cracking caused by restraint to the load-independent shrinkage and temperature deformations. In addition, the final deflection of a slab depends very much on the extent of initial cracking which, in turn, depends on the construction procedure (shoring and re-shoring), the amount of early shrinkage, the temperature gradients in the first few weeks after casting, the degree of curing and so on. Many of these parameters are, to a large extent, out of the control of the designer. In field measurements of the deflection of many identical slab panels [3,4] a large variability was reported. Deflections of identical panels after one year differed by over 100% in some cases. These differences can be attributed to the different conditions (both in terms of load and environment) that existed in the first few weeks after the casting of each slab.

Over the past several years, numerous cases have been reported, in Australia and elsewhere, of flexural elements for which the deflection calculated in accordance with AS3600 [1] is far less than the actual deflection. Many of these elements complied with the code, but still deflected excessively. Evidently, the deflection calculation procedures specified in the code do not adequately model the in-service behaviour of some beams and slabs.

The need for a more reliable deflection calculation procedure has been exacerbated by the introduction in recent years of higher strength reinforcing steels. The use of 500 MPa steel, rather than 400 MPa steel, for example, results in less steel being required for strength and, consequently, less stiffness being available after cracking, leading to greater deflection and crack widths under service loads. In the design for serviceability, attention must be given to the specification of (i) an appropriate concrete mix, with suitably low creep and shrinkage characteristics, and (ii) a suitable construction procedure, involving acceptably long stripping times, adequate propping, effective curing and rigorous on-site supervision.

In this paper, the existing simplified approaches for deflection calculation in AS3600 are presented, their deficiencies are illustrated and discussed and suggestions for improving the procedures are proposed and evaluated. The corresponding procedures in EC2 [2] are also evaluated.

EXISTING PROCEDURE IN AS3600-1994

Immediate Deflection

The simplified approach for the calculation of deflection of a beam in AS3600-1994 is similar to that in ACI318-1995. An effective second moment of area, I_{ef}, for the member is used to calculate immediate deflection, together with the value of the elastic modulus for concrete, E_c, at the time of first loading. The value of I_{ef} for the member is determined from the value of I_{ef} at nominated cross-sections. At each of these cross-sections, I_{ef} is given by (Branson, [5])

$$I_{ef} = I_{cr} + (I_g - I_{cr})(M_{cr} / M_s)^3 \quad \leq I_g \qquad (1)$$

where I_{cr} is the second moment of area of the fully-cracked section (calculated using modular ratio theory); I_g is the second moment of area of the gross concrete section about its centroidal axis; M_s is the maximum bending moment on the section caused by the short-term serviceability load or the construction load; M_{cr} ($= Z\, f'_{cf}$) is the cracking moment at the section; Z is the section modulus of the uncracked section (referred to the extreme tensile fibre); and f'_{cf} ($= 0.6\sqrt{f'_c}$) is the characteristic flexural tensile strength of the concrete.

The immediate deflection of a slab subjected to uniformly distributed loads may be calculated by applying the above procedures to an equivalent slab strip (or beam) defined as follows:

- For a one-way slab, a prismatic slab strip (beam) of unit width;
- For a two-way edge supported slab, a prismatic slab strip of unit width through the centre of the slab, spanning in the short direction, L_x, with the same conditions of continuity as the slab in that direction and with the load distributed so that the proportion of load carried by the slab strip is given by $L_y^4 / (\delta L_x^4 + L_y^4)$. The term δ depends on the support conditions of the slab edges [1]. The deflection of this slab strip is greater than the deflection of the slab (by about 30% for a square panel and decreasing to close to zero for a rectangular panel with $L_y / L_x \geq 2$) due to the torsional stiffness of the two-way slab.
- For two-way flat slabs having multiple spans, the design strips of the idealised frame.

Time-Dependent Deflection

In the absence of more accurate calculations, that part of the deflection which occurs after the immediate deflection, and is the result of creep and shrinkage of the concrete, may be calculated by multiplying the immediate deflection caused by the sustained load by a multiplier, k_{cs}, given by

$$k_{cs} = [2 - 1.2(A_{sc}/A_{st})] \geq 0.8 \qquad (2)$$

where A_{sc}/A_{st} is the ratio of compressive to tensile steel at midspan for a simply-supported or continuous span, and at the support for a cantilever.

DEFLECTION CALCULATIONS IN EC2

The deflection of a flexural member may be calculated from the curvature at selected points along the member by integration. In the determination of the load-dependent deflection of a cracked member, an effective curvature κ is used which is intermediate between the curvatures κ_I and κ_{II} determined on the basis of the uncracked and the fully-cracked condition, respectively, and is specified as

$$\kappa = \zeta\, \kappa_{II} + (1-\zeta)\kappa_I \tag{3}$$

where $\kappa_I = M / E_{ef} I_g$ and $\kappa_{II} = M / E_{ef} I_{cr}$. The distribution coefficient ζ is given by

$$\zeta = 1 - \beta_1 \beta_2 (M_{cr} / M)^2 \tag{4}$$

where $\beta_1 = 1$ for high bond (deformed) bars and 0.5 for plain bars; $\beta_2 = 1$ for a single short-term load and 0.5 for sustained loads or for many cycles of repeated loading; E_{ef} is the effective modulus of the concrete and may be taken as E_{cm} for short-term loading and, for long-term loading, $E_{ef} = E_{cm} /(1 + \phi_c)$; E_{cm} and ϕ_c are the elastic modulus of concrete (see Table 1) and the creep coefficient, respectively; M_{cr} is the cracking moment required to cause an extreme fibre tensile stress equal to the mean tensile strength of concrete f_{ctm} (see Table 1); and M is the in-service moment on the section due to the loading under consideration. It should be noted that the use of E_{ef} overestimates the creep induced curvature on a cracked section, where the change in curvature due to creep is not directly proportional to the change in strain in the compressive concrete fibres.

In the determination of ζ, it is prudent to take M as the maximum likely in-service moment at the section under consideration.

The load-independent curvature caused by shrinkage κ_{cs} on a cracked section should be added to the load-dependent curvature κ and is given by

$$\kappa_{cs} = \varepsilon_{cs}\, n_{ef}\, S / I \tag{5}$$

where ε_{cs} is the free shrinkage strain; n_{ef} is the effective modular ratio (E_s / E_{ef}); S is the first moment of the reinforcement about the centroid of the section; and I is the second moment of area of the section. S and I should be calculated for both the uncracked and the fully-cracked condition and Eq. 5 used to calculate $(\kappa_{cs})_I$ and $(\kappa_{cs})_{II}$, respectively. The effective (or average) shrinkage-induced curvature κ_{cs} is obtained using the distribution coefficient ζ and an equation similar in form to Eq. 3.

Table 1 Concrete Properties EC2 [2]

f'_c (MPa)	20	25	30	35	40	45	50
E_{cm} (GPa)	29	30.5	32	33.5	35	36	37
f_{ctm} (MPa)	2.2	2.6	2.9	3.2	3.5	3.8	4.1

DEFICIENCIES OF AS3600 PROCEDURE

Eq. 1 provides a reasonable estimate of the stiffness of reinforced concrete flexural members with tensile reinforcement ratios typically greater than 0.6% and when M_s is significantly greater than M_{cr}. In these cases, I_{ef} is not much greater than I_{cr} and tension stiffening is not significant. However, in the case of lightly reinforced beams, and most slabs, M_s may be close to or even less than M_{cr}. In such cases, the calculated value of I_{ef} is close to, or perhaps equal to, the second moment of area of the uncracked section I_g and may overestimate stiffness considerably. The loss of stiffness that will inevitably occur due to cracking is not accounted for in the deflection calculations. Shrinkage induced tension and tension caused by temperature changes can reduce M_{cr} from the value specified above. EC2 overcomes this deficiency to some extent by the inclusion of $\beta_2 = 0.5$ in Eq 4 for long-term loading.

It is unwise in deflection calculations to assume that flexural members are uncracked or that tension stiffening will cause I_{ef} to remain close to I_g, particularly under repeated loading and for long-term deflection calculations. In addition to causing tension in uncracked regions that may result in time-dependent cracking, shrinkage also induces tension between the cracks in cracked regions causing a gradual breakdown of tension stiffening with time.

The use of the deflection multiplier k_{cs} to calculate time-dependent deflections is simple and convenient and, provided the section is initially cracked under short term loads, usually provides a 'ball-park' estimate of deflection. When used for lightly stressed elements, however, it is misleading and grossly inadequate. The approach combines the load independent shrinkage deflection with the load-dependent deflection. It ignores the creep and shrinkage characteristics of the concrete, the environment, and the age at first loading. At best, it provides a very approximate estimate. The long-term creep and shrinkage deflection should be calculated separately. Well establish methods are available for calculating the time-dependent behaviour of a reinforced concrete cross-section [7].

To illustrate the problems, one of a number of simply-supported one-way slabs tested under a constant sustained loading for a period of 260 days is considered. The slab was prismatic with cross-section 1000 mm wide and 180 mm thick and contained tensile reinforcement consisting of six – 12 mm diameter bars ($A_{st} = 660$ mm^2) at an effective depth $d = 150$ mm. The slab span was 5 m and the sustained load was 2 kPa, in addition to its self-weight (ie. w_{sus} = 6.4 kPa). It is noted that the specimen satisfies the requirements of AS3600 for strength.

The slab was moist cured for 14 days after casting at which time the sustained load was fist applied and testing commenced. The characteristic concrete compressive strength at 28 days was determined from cylinder tests to be 31 MPa. The shrinkage strain in the concrete, ε_{cs}, was measured on companion specimens of the same cross-sectional dimensions as the slab and with the same curing conditions. The creep coefficient, ϕ_c, was measured on cylinders under a constant sustained compressive stress of 5.0 MPa, first applied at age 14 days. After 260 days, the measured values were $\varepsilon_{cs} = 690$ με and $\phi_c = 2.6$.

Immediately after the sustained load was applied to the specimen, the deflection at midspan was 3.9 mm and no evidence of cracking was observed. The first cracks in the slab were noticed on the slab soffit after 6 days under load and over the next 45 days additional cracks developed. No new cracks were recorded after 51 days under load, but the width of the cracks increased throughout the period of testing. The deflection at midspan after 260 days under load was 28.9 mm (with only 2.1mm being recorded in the last 100 days). With a span to final deflection ratio of 173, the slab is well outside the code's minimum allowable span to final deflection ratio of 250 and would be unserviceable for most applications.

With $f'_c = 31$ MPa, the calculated values of elastic modulus and flexural tensile strength of concrete (in accordance with AS3600) are $E_c = 28120$ MPa and $f'_{cf} = 3.34$ MPa, respectively. For this specimen, the moment at midspan $M_s = 20.0$ kNm; the cracking moment $M_{cr} = 18.04$ kNm; the reinforcement ratio $p = 0.0044$; the modular ratio $n = 7.11$; and the second moment of area of the gross section and of the cracked section are $I_g = 486 \times 10^6$ mm^4 and $I_{cr} = 76.2 \times 10^6$ mm^4, respectively. From Eq.1, the effective second moment of area is $I_{ef} = 377 \times 10^6$ mm^4 and, from Eq 2, the long-term multiplier $k_{cs} = 2.0$.

The calculated immediate deflection of the simply-supported slab is $\Delta_i = 4.9$ mm and the final deflection is $\Delta = (1 + k_{cs})\Delta_i = 14.7$ mm $= L/339$.

The procedure predicts only about 50% of the actual final deflection of this slab and with a predicted time-dependent deflection of only 9.8 mm (= $L/510$), a designer may even be fooled into accepting this slab as suitable for supporting brittle partitions. Clearly, the simplified procedure in AS3600-1994 for calculating deflections is inadequate.

PROPOSED IMPROVEMENTS

The Standards Australia Working Group BD/2/7 is currently considering several alternatives for improving the reliability of the provisions relating to simplified deflection calculations in AS3600. Two of the alternatives are presented and evaluated here.

Alternative 1

To account for the breakdown of tension stiffening under long-term or cyclic loads, the following modification to Branson's equation is suggested:

$$I_{ef} = I_{cr} + (I - I_{cr})\ \beta\ (M_{cr} / M_s)^3 \ \leq I \tag{6}$$

where $\beta = 1.0$ for immediate deflection and 0.4 for long-term or cyclic loading.

The curvature due to the sustained, in-service load is calculated as $M_s / E_{cm} I_{ef}$, where I_{ef} is obtained from Eq 6 using $\beta = 0.4$ and E_{cm} is a modified effective modulus for concrete, which may be taken as

$$E_{cm} = E_c / (1 + \phi_c / \alpha) \tag{7}$$

The factor α accounts for the degree of cracking and the 'braking' action of the reinforcement and varies between α_I for an uncracked section and α_{II} for a fully-cracked section and may be taken as

$$\alpha = \alpha_{II} - (\alpha_{II} - \alpha_I)\ \beta\ (M_{cr} / M_s)^3 \tag{8}$$

For a singly-reinforced section, α_I is typically in the range 1.0 to 1.6 and α_{II} is in the range 4 to 10, both depending on the amount of tensile steel. Eqs. 9a and 9b are empirical estimates of α_I and α_{II} [6].

$$\alpha_I = 1.0 + [45p - 900p^2][1 + A_{sc}/A_{st}] \tag{9a}$$

$$\alpha_{II} = [20000\,p^2 - 700\,p + 9.0][1 + 0.7A_{sc}/A_{st}] \tag{9b}$$

where A_{st} is the area of bonded reinforcement in the tensile zone and $p = A_{st}/bd$.
The shrinkage induced curvature κ_{cs} on a cracked reinforced concrete section can be taken as

$$\kappa_{cs} = (\kappa_{cs})_{II} - [(\kappa_{cs})_{II} - (\kappa_{cs})_I]\ \beta\ (M_{cr}/M_s)^3 \tag{10}$$

where $(\kappa_{cs})_I$ and $(\kappa_{cs})_{II}$ are the corresponding curvatures on an uncracked and a fully cracked cross-section, respectively and are approximated by [6]

$$(\kappa_{cs})_I = \left[\frac{0.7\varepsilon_{cs}}{D}\right]\left[1 - \frac{A_{sc}}{A_{st}}\right] \quad \text{and} \quad (\kappa_{cs})_{II} = \left[\frac{1.2\varepsilon_{cs}}{D}\right]\left[1 - \frac{A_{sc}}{A_{st}}\right] \tag{11}$$

Further, to account for the fact that M_{cr} may be substantially reduced by load independent tension, it is recommended that whenever $M_s < M_{cr}$, the ratio of M_s/M_{cr} should be taken as 1.0 in each of Eqs. 6, 8 and 10. In this way, cracking due to shrinkage and temperature changes will always be accounted for in deflection calculations even though the level of moment due to service loads may not be sufficient, of itself, to cause cracking.

When the load-induced and shrinkage-induced curvatures are calculated at selected sections along the beam and summed, the deflection may be obtained by double integration. For a reinforced concrete continuous span, with the degree of cracking varying along the span, the curvature at the left and right supports, κ_L and κ_R, and the curvature at midspan, κ_M, may be calculated at any time after loading and the deflection at midspan, Δ, may be approximated by assuming a parabolic curvature diagram along the span, L:

$$\Delta = \frac{L^2}{96}(\kappa_L + 10\kappa_M + \kappa_R) \tag{12}$$

The above equation will give a reasonable estimate of deflection even when the curvature diagram is not parabolic and is a useful expression for use in deflection calculations.

Alternative 2

An alternative and perhaps more rational approach is to include the shrinkage induced tension in the estimation of the cracking moment. It can be easily shown [6] that the tension induced by shrinkage in the extreme tensile fibre of an uncracked singly reinforced concrete cross-section may be approximated by

$$f_{cs} = 3.5\bar{p}\ \bar{n}\varepsilon^*_{cs}\ \bar{E}_{ef}\ /(1 + 3\bar{p}\ \bar{n}) \tag{13}$$

where $\overline{E}_{ef}$ is the age-adjusted effective modulus for concrete and may be taken as $\overline{E}_{ef} = E_c/(1+0.8\phi_c^*)$; $\overline{p} = A_{st}/A$; $\overline{n} = E_s/\overline{E}_{ef}$; ε_{cs}^* and ϕ_c^* are the final long-term values of shrinkage strain and the creep coefficient; and A is the gross area of the concrete section. The above expression assumes that the overall depth of the section is 1.15d. By assuming conservative values for the elastic modulus and the creep coefficient of concrete, f_{cs} may be simplified to

$$f_{cs} = 3.5\overline{p}E_s\varepsilon_{cs}^*/(1+50\overline{p}) \tag{14}$$

The shrinkage induced tension, f_{cs} , will develop in an uncracked section and the moment required to cause cracking M_{cr} will produce an extreme fibre tensile stress of (f'_{cf} - f_{cs}). For a reinforced concrete beam section, therefore

$$M_{cr} = Z\,(f'_{cf} - f_{cs}) \qquad \text{but} \quad \geq 0.0 \tag{15}$$

With this adjustment, Eq. 1 may be used to calculate I_{ef} and this is then used to recalculate the curvature due to the sustained long-term load, $M_s/E_{cm}I_{ef}$ (as for Alternative 1). The shrinkage-induced curvature is obtained from Eq. 10 and the final deflection from Eq. 12. In the determination of α (Eq. 8) and κ_{cs} (Eq. 10), a value of $\beta = 1$ should be used since the effects of shrinkage induced tension have been accounted for in the reduced value of M_{cr}.

EVALUATION OF PROPOSALS

The measured deflection of the one-way slab described earlier is here used to evaluate the proposed improvements to the simplified deflection calculation procedures.

Alternative 1

The calculated immediate deflection Δ_i = 4.9 mm was determined previously. The value of I_{ef} used to calculate the long-term deflection under the sustained load is obtained from Eq. 3:

$$I_{ef} = [76.2 + (486 - 76.2) \times 0.4 \times (18.04/20.0)^3] \times 10^6 = 196.5 \times 10^6 \text{ mm}^4$$

With $p = A_{st}/bd$ = 0.0044, Eq 9 gives α_I = 1.18 and α_{II} = 6.31, and, from Eq. 8, $\alpha = 4.80$. Using the measured creep coefficient ϕ_c = 2.6, the modified effective modulus for concrete is obtained from Eq. 7:

$$E_{cm} = 28120/(1+2.6/4.80) = 18240 \text{ MPa}$$

and the curvature due to the sustained load is then

$$\kappa = M_s/(E_{cm}I_{ef}) = 20\times10^6/(18240\times196.6\times10^6) = 5.58\times10^{-6}\text{ mm}^{-1}$$

Due to the final shrinkage strain of ε_{cs} = 690 με, Eq 11 gives $(\kappa_{cs})_I = 2.68 \times 10^{-6}\,\text{mm}^{-1}$ and $(\kappa_{cs})_{II} = 4.60 \times 10^{-6}\,\text{mm}^{-1}$. At midspan, the shrinkage induced curvature is (Eq. 10), $\kappa_{cs} = 4.04 \times 10^{-6}\,\text{mm}^{-1}$ and, at each support, $\kappa_{cs} = 2.68 \times 10^{-6}\,\text{mm}^{-1}$.

The final long-term deflection is calculated from Eq. 12, in which $\kappa_L = \kappa_R = 2.68 \times 10^{-6}\,\text{mm}^{-1}$ and $\kappa_M = (5.58 + 4.04) \times 10^{-6} = 9.62 \times 10^{-6}\,\text{mm}^{-1}$. That is

$$\Delta = \frac{5000^2}{96}(2.68 + 10 \times 9.62 + 2.68) \times 10^{-6} = 26.5\,\text{mm}.$$

This underestimates the measured deflection (28.9 mm) by 8%. Considering the variability of the material properties used in the calculations, the agreement between calculated and measured deflection is considered to be good. Alternative 1 is relatively simple to use and gives reliable estimates of deflection, but in members with low tensile steel quantities and when M_s is not much larger than M_{cr} (such as the slab considered here), the method slightly underestimates actual deflections.

Alternative 2

As for Alternative 1, the calculated immediate deflection is Δ_i = 4.9 mm. Using the measured creep coefficient ϕ_c = 2.6 and shrinkage strain ε_{sc} = 690 με, the age-adjusted effective modulus is $\overline{E}_e = E_c/(1 + 0.8\phi_c) = 9130\,\text{MPa}$. The age-adjusted modular ratio is $\overline{n}$ = 21.9 and $\overline{p}$ = 0.00367. The shrinkage induced tension in the concrete is (Eq. 13): f_{cs} = 1.43 MPa (or alternatively, using the simplified Eq. 14, f_{cs} = 1.50 MPa)

The cracking moment and the effective moment of inertia are calculated from Eqs. 15 and 1, respectively:

$$M_{cr} = \frac{1000 \times 180^2}{6}(3.34 - 1.43) = 10.31\ \text{kNm}.$$

$$I_{ef} = [76.2 + (486 - 76.2) \times (10.31/20.0)^3] \times 10^6 = 132.4 \times 10^6\ \text{mm}^4.$$

From Eq. 8 (with $\beta = 1$), $\alpha = 5.61$ and, from Eq. 7, $E_{cm} = 19210\,\text{MPa}$. Therefore, the final curvature at midspan produced by the sustained load $(\kappa)_M = M_s/(E_{cm} I_{ef}) = 7.86 \times 10^{-6}\ \text{mm}^{-1}$.

As for Alternative 1, $(\kappa_{cs})_I = 2.68 \times 10^{-6}\,\text{mm}^{-1}$ and $(\kappa_{cs})_{II} = 4.60 \times 10^{-6}\,\text{mm}^{-1}$. The shrinkage curvature at midspan is estimated (Eq.10 with $\beta = 1$) to be $(\kappa_{sc})_M = 4.37 \times 10^{-6}\ \text{mm}^{-1}$.

The total curvature at midspan is κ_M = (7.86 + 4.37) x 10^{-6} = 12.23 x 10^{-6} mm^{-1} and at the supports, κ_L = κ_R = 2.68 x 10^{-6} mm^{-1}. From Eq 12, the final deflection at midspan is $\Delta = 33.2$ mm, which is a conservative estimate of the final deflection.

The advantages that these procedures have over the use of the deflection multiplier, k_{cs}, are (i) the slab behaviour is modelled more realistically and more reliably; and

(ii) a designer can take advantage of the significant benefits that can be realised with regard to deflection behaviour by using better quality concretes in flexural members.

Eurocode 2 Procedure

From Table 3.1, for $f'_c = 31$ MPa, E_{cm} = 32300 MPa and f_{ctm} = 2.96 MPa. The cracking moment is M_{cr} = 16.0 kNm and the effective modulus is E_{ef} = 8970 MPa. The effective modular ratio is n_{ef} = 22.3 and the corresponding second moment of the cracked section is I_{cr} = 189 x 10^6 mm^4. The depth to the centroidal axis of the cracked section x = 53.3 mm.

The curvatures of the uncracked and fully-cracked cross-sections at midspan are $\kappa_I = M / E_{ef} I_g = 4.59 \times 10^{-6}$ mm^{-1} and $\kappa_{II} = M / E_{ef} I_{cr} = 11.80 \times 10^{-6}$ mm^{-1}. From Eq 4, ζ = 0.68 and the effective load dependent curvature at midspan (Eq. 3) is $\kappa = 9.49 \times 10^{-6}$ mm^{-1}.

The shrinkage induced curvatures for the uncracked and fully-cracked condition are (Eq. 5) $(\kappa_{cs})_I = 1.89 \times 10^{-6}$ mm^{-1} and $(\kappa_{cs})_{II} = 5.22 \times 10^{-6}$ mm^{-1} and at midspan (Eq. 3), κ_{cs} = 4.20×10^{-6} mm^{-1}.

The final curvature at midspan is therefore $\kappa_M = 13.69 \times 10^{-6}$ mm^{-1} and at each support κ_L = $\kappa_R = 1.89 \times 10^{-6}$ mm^{-1}. From Eq. 12, the midspan deflection is $\Delta = 36.6$ mm.

The Eurocode 2 approach overestimates creep deflection and, hence, final deflection.

CONCLUSIONS

The inadequacies of the simplified procedures in AS3600-1994 [1] for deflection calculations of reinforced concrete beams and slabs specified have been discussed and illustrated by comparison of deflection calculations with the measured long-term deflection of a one-way slab. For lightly reinforced beams and most slabs, the procedures in the Standard fail to adequately model the loss of stiffness caused by time-dependent cracking and the gradual reduction in tension stiffening caused by shrinkage. As a consequence, the calculated deflection, almost invariably, underestimates the actual deflection of such members.

In contrast, the procedure in Eurocode 2 [2] tends to overestimate the effects of creep in a cracked region by adopting the effective modulus for concrete, thereby assuming incorrectly that the creep induced curvature is the product of the initial curvature and the creep coefficient.

Two alternative methods for improving the calculation procedure, which are currently under consideration by Standards Australia, have been presented and evaluated. Both alternatives provide deflection calculations which more closely agree with measured deflections.

REFERENCES

1. AS3600-1994, Australian Standard for Concrete Structures, Standards Australia, 1994.

2. EUROCODE 2: PART 1, Design of Concrete Structures, DD ENV 1992-1-1, 1992.

3. SBAROUNIS, J.A., Multi-storey Flat Plate Buildings – Measured and Computer One-year Deflections, *Concrete International*, Vol 6, No.8, August 1984, pp. 31-35.

4. JOKINEN, E.P. and SCANLON, A., Field Measured Two-way Slab Deflections, CSCE Annual Conference, Saskatoon, Canada, May 1985, 16pp.

5. BRANSON, D.E., Instantaneous and Time-Dependent Deflection of Simple and Continuous RC Beams, Alabama Highway Res. Rpt, No.7, Bur. of Public Roads, 1963.

6. GILBERT, R.I., Serviceability Considerations and Requirements for High Performance Reinforced Concrete Slabs, Proceedings Int. Conf. On High Performance High Strength Concrete, Curtin University of Technology, Perth, August 10-12 1998, pp 425-439.

7. GILBERT, R.I., Time Effects in Concrete Structures, Elsevier , Amsterdam, 1988, 321p.

VIBRATION OF FRESH CONCRETE: EXPERIMENTAL SET-UP AND PRELIMINARY RESULTS

M N Soutsos

J H Bungey

M J Brizell

University of Liverpool

United Kingdom

ABSTRACT. Electronic equipment known as accelerometers have been used in this research project to measure the sinusoidal wave form created by mechanical vibrators and thus gain a better understanding of the consolidation process. The amplitude of the sinusoidal signal was calculated from the acceleration measured at distances up to 250mm from the surface of the poker vibrator. Preliminary tests indicate that the amplitude of the vibratory wave decays exponentially with distance. The damping coefficient is greater for superplasticized high strength concrete mixes with low water-cement ratios than it is for normal strength concrete mixes. This means that the radius of influence of a poker vibrator is significantly smaller for high strength concretes as compared to normal strength concretes. The rheological properties yield (g) and plastic viscosity (h) values determined from tests carried out with Tattersall's two point test apparatus appear to be related to the damping coefficients of concrete mixes with equal paste content.

Keywords: Acceleration, Amplitude, Compaction, Concrete vibrators, Consolidation, Damping coefficient, High strength concrete, Plastic viscosity (h) value, Rheology, Slump value, Tattersall's two point test, Vibratory wave, Workability, Yield (g) value.

Dr M N Soutsos received his BEng and PhD degrees from the University College London. He is currently a lecturer in the Department of Civil Engineering, The University of Liverpool. His research interests include: cement replacement materials, thermal stresses in concrete, nondestructive testing, as well as mix design and rheological properties of high strength concrete.

Professor J H Bungey holds a MSc degree in concrete structures and technology from Imperial College, London and a PhD from Liverpool University where he is currently Head of the Civil Engineering Department. His research interests encompass concrete materials, design and performance including non-destructive testing and he has authored numerous papers and books on these topics.

Mr M J Brizell received his BEng degree from the Department of Civil Engineering, The University of Liverpool, in 1996 and where he is currently a postgraduate research student carrying out research on high performance concrete and its behaviour during vibration.

INTRODUCTION

Consolidation, also called compaction, is the process of removing entrapped air from freshly placed concrete. Several methods and techniques are available; the choice depending mainly on the workability of the mixture, placing conditions, and degree of air removal desired. The most widely used consolidation method is vibration which consists of subjecting freshly placed concrete to rapid vibratory impulses which liquefy the mortar and drastically reduce the internal friction between aggregate particles. Consolidation by vibration is best described as consisting of two stages; the first comprising subsidence or slumping of the concrete, and the second a deaeration (removal of entrapped air bubbles). The two stages may occur simultaneously, with the second stage under way near the vibrator before the first stage has been completed at greater distances. Concrete vibrators can be divided into two main classes: internal and external. The internal types, which are the subject of the research project described in this paper, generally use a rotating unbalanced weight or eccentric inside a vibrator casing. They therefore generate harmonic motion, characterised by a sinusoidal wave form. Evidence strongly indicates that the effectiveness of an internal vibrator depends mainly on the head diameter, frequently, and amplitude. The radius of action, and hence the insertion spacing, depends not only on the characteristics of the vibrator, but also on the workability of the mixture. According to ACI Committee 309(1): "there is no simple method for determining the amplitude of a vibrator in concrete". A technique for determining the sinusoidal wave transmission through fresh concrete has been developed at The University of Liverpool and is being used to develop a better understanding of the interaction of the concrete vibrators with fresh concrete.

The aims of the research described in this paper, in line with the above, are as follows:

(i) To express workability according to classical rheological models, such as the Bingham, with tests carried out using Tattersall's two point test.

(ii) To relate the rheological constants obtained in (i) to the behaviour of concrete during the second stage of the vibration process, i.e. the de-aeration stage (removal of entrapped air).

MATERIALS, EXPERIMENTAL PROCEDURES AND EQUIPMENT

Concrete Mixes

A total of 15 different mixes grouped into 4 separate series with various values of water-cement ratios ranging between 0.60 to 0.26 were tested. Normal strength concrete mixes were designed according to the guidelines published by the U.K. Department of the Environment(2) and are referred to as the BRE mixes. The concrete constituent proportions are given in Table 1. Superplasticized high strength concrete mixes were designed on a volumetric basis using the Modified Maximum Density Theory described in detail elsewhere(3,4) and are referred to as the MMDT mixes. The fine aggregate proportion (37.0%) was that required to give the lowest binder content for any given workability. The paste (i.e. cement + water) volume was that required to fill the voids in the combined aggregate system (27.5% by volume) plus an extra amount of overfill equivalent to 5% of the total concrete volume. This overfill level was that which gives the highest slump for any particular cement content. The resulting mix proportions for each of the water-cement ratios are given in Table 1.

Table 1 Concrete mixes

MIX.	W/C	MATERIAL PROPORTIONS (kg/m^3)				SUPER-PLASTICIZER (% solids by wt. of cement) ****	SLUMP (mm)
		Water	Cement *	Sand **	10mm Gravel ***		
1-1A	0.60	225	375	860	860		70
1-2A	0.55	210	382	910	860		50
1-3A	0.50	225	450	840	840		55
1-4A	0.45	200	440	820	920		60
2-1B	0.60	225	375	860	860		75
2-2B	0.50	225	450	840	840		55
2-3B	0.45	200	440	820	920		60
2-4B	0.40	200	500	750	900		30
3-1A	0.40	156	390	636	1083	0.3	70
3-1A	0.38	153	403	636	1083	0.5	80
3-3A	0.35	147	421	636	1083	0.6	75
3-4A	0.32	141	441	636	1083	0.8	85
4-1B	0.42	160	380	636	1083	0.3	55
4-2B	0.38	153	403	636	1083	0.5	60
4-3B	0.32	141	441	636	1083	1.1	50

* typical normal Portland (Type I) cement
** natural sand from Hope quarry with a fineness modulus of 2.10
*** Staffordshire Pink gravel aggregate with maximum aggregate size of 10 mm
**** a sulphonated napthalene formaldehyde condensate superplasticizer with an active solids content of 50% by weight

NOTE:

- all "A" mixes have been tested in the Tattersall's two point apparatus
- all "B" mixes have been used to obtain the decay of amplitude of sinusoidal waves with distance from the poker vibrator surface

Mixing Procedure

Mixing was carried out in a laboratory pan mixer. Materials were added in the order: coarse aggregate, cement, sand and water. The superplasticizer was premixed with the water to ensure consistency of action throughout the test programme. Mixing was then started and was continued for 5 minutes after all the materials had been added. The mixing time was sufficient to allow for absorption of the water by the aggregate. The slump was measured as soon as practical after mixing was completed. This avoided the effects of any rapid loss of slump often associated with high strength mixtures. The concrete was subsequently either tested in the two point apparatus or in the accelerometer-supporting test-rig. Both tests took less than 10 minutes, after which the concrete was returned to the mixer, remixed for a further

Figure 1 Tattersall's two point test apparatus

one minute, and again tested for slump. The average of the two slump values, i.e. before and after the test, was calculated. The two slump values were found in most cases not to differ by more than 30 mm.

Tattersall Two-Point Test Apparatus

The concrete under test in the Tattersall Two-Point Test Apparatus, shown in Figure 1, is contained within a cylindrical bowl and is sheared by a suitable impeller which is driven by an electric motor operating through an infinitely variable hydraulic transmission and a reduction gear. The pressure developed in the oil in the hydraulic transmission is measured by a Budenburg pressure gauge. The pressure produced by concrete shearing is obtained from the total pressure by subtracting the pressure produced by machine idling. This net value may be converted into impeller torque after calibration. The impeller torque T and speed N are found to be related by the linear equation:

$$T = g + hN$$

By measuring the torque produced on an impeller rotating in fresh concrete at various speed settings, the values of (g) and (h) can be determined, which therefore define the workability of a particular concrete mix. Different concretes at different workabilities produce different

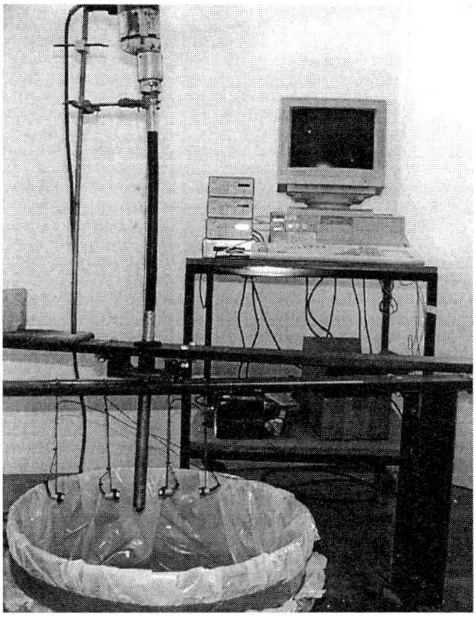

Figure 2 Accelerometer-supporting test-rig

values of (g) and (h). The resemblance of the equation to the rheological equation of Bingham flow is often noted (Bingham: $\tau = \tau_0 + \mu\gamma$ where τ = shear stress; τ_0 = yield value; μ = plastic viscosity, γ = shear rate). Theoretical justification that fresh concrete approximates to the Bingham model and that the two constants (g) and (h) are directly proportional to the Bingham constants τ_0 and μ has been provided by Tattersall(5).

Accelerometer Set-Up

The accelerometer supporting test rig is shown in Figure 2. The poker vibrator used was a pistol grip type. The diameter of the head was 27.5mm, and it had an amplitude and frequency in air of 0.6mm and 283Hz respectively.

Eight piezo-electric accelerometers were buried in the concrete at a depth of 150mm and at distances of up to 250mm from the poker vibrator. Each accelerometer was approximately equal in size and weight to the larger stones used in the concrete mixtures. It was essential that these devices embedded in the concrete should not interfere with the propagation of the vibratory waves. They were supported using hinged stainless steel rods and were arranged to record horizontal vibrations. Their outputs were fed into electronic equipment that enabled them to be recorded by a Personal Computer. These could then be converted to acceleration values (mm/sec^2) and using the standard equations for sinusoidal oscillation the peak amplitude of the wave was calculated.

RESULTS AND DISCUSSION

Tattersall's Two-Point Tests

Slump values of 55 ± 20 mm and 150 ± 30mm were targeted for the two series of mixtures that were tested using Tattersall's Two Point Test Apparatus. Impeller speed versus torque plots were used to obtain values for the yield (g) and viscosity (h) values that are shown in Figures 3 and 4 respectively. It appears that for the BRE mixes, both the yield (g) and plastic viscosity (h) values, increase slightly with decreasing water-cement ratios. Small variations in workability, i.e. slump values, appear to have a bigger effect on the yield (g) and plastic viscosity (h) values than changes in the water-cement ratio. Indeed, mixes with slump values in the range 45 to 75 mm had much higher yield (g) values as compared to the higher workability mixes of the second series.

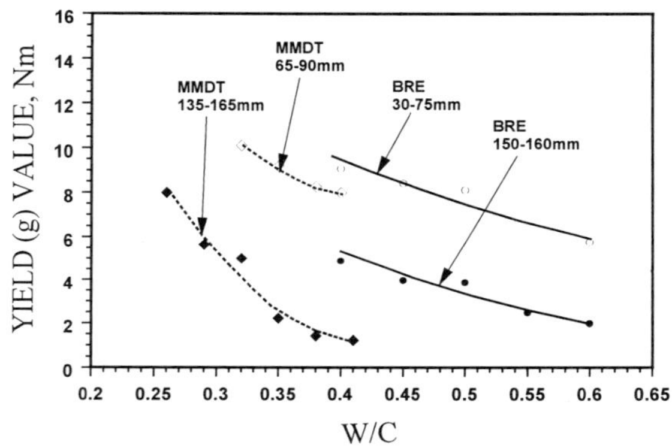

Figure 3 Yield value (g) versus water-cement ratio

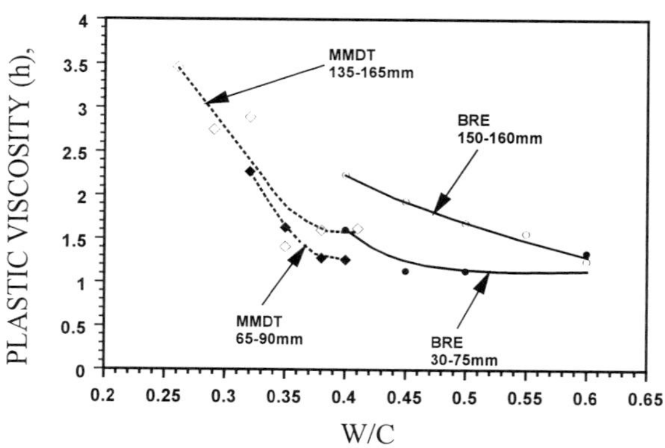

Figure 4 Plastic viscosity (h) value versus water-cement ratio

The superplasticized high strength concrete mixes, behaved differently from the normal mixes. Two point test results have shown, that as the water-cement ratio was lowered, the yield (g) and plastic viscosity (h) values increased considerably despite the mixes all having a slump of 150 mm. The addition of a superplasticizer to a mix with a fixed water-cement ratio is known to mainly reduce the yield stress, i.e. its (g) value, without significantly affecting the plastic viscosity, i.e. its (h) value. Decreasing the water-cement ratio increases both the yield (g) and plastic viscosity (h) values. The addition of a superplasticizer would therefore be expected to help decrease the rate at which the yield stress increases as the water-cement ratio is reduced, but not reduce the increase in the plastic viscosity (h) value. The increase in the yield (g) value at constant slump was therefore somewhat unexpected. It seems to arise from the inherently higher solids content of the paste used as the water-cement ratio was lowered.

Decay Of The Amplitude Of Sinusoidal Waves In Concrete

The amplitude of the sinusoidal vibratory wave has been found to decay exponentially with distance from the poker vibrator, see Figure 5. This is in agreement with Dessoff's postulated formula(6) for the geometrical energy distribution due to the radial generation of compression waves:

$$s_2 = s_1 \sqrt{\frac{\mathbf{R}_1}{\mathbf{R}_2}} \mathbf{e}^{-(\Omega/2)(\mathbf{R}_2 - \mathbf{R}_1)}$$

where Ω is the coefficient of damping, and $\mathbf{s}_1$ and $\mathbf{s}_2$ are the amplitudes at the respective distances of $\mathbf{R}_1$ and $\mathbf{R}_2$ from the center of the internal vibrator. The above equation can be rewritten, by taking logs, as:

$$\ln\left(s_2 \sqrt{\mathbf{R}_2}\right) = \mathbf{k} - \frac{\Omega}{2}\mathbf{R}_2$$

where k is a constant. The damping coefficient Ω can then be obtained from a plot of

$$\ln\left(s_2 \sqrt{\mathbf{R}_2}\right) \text{ versus } \mathbf{R}_2.$$

The amplitude of the poker vibrator was found to decrease from 0.6mm in air to around 0.47 ± 0.01mm when placed in concrete. Similar reductions in amplitude, i.e. 25 to 30%, have been reported by ACI Committee 309(7). It is apparent from Figure 5 that the amplitude generated around the vibrating head is not the same as the amplitude of the vibrator. It appears that during vibration, some of the cement paste separates from the aggregates and surrounds the vibrating head. It subsequently cannot get pushed by the vibrator back into the concrete but instead moves around the vibrating head to areas of low pressure created by the sinusoidal waves during half of their periods. This results in a reduction of the energy transmission from the vibrator to the fresh concrete. According to ACI Committee 309(7) this reduction is complimented, at increased vibration intensities, by the creation of vapor and gas bubbles which act as shock absorbers and therefore dampen the compression waves even further. This phenomenon has been termed *"cavitation"*. Its effects are shown in Figure 6. The acceleration recorded at a distance of 10mm from the poker head starts at zero and increases to a maximum value of 285m/sec^2 after the vibrator is switched on. However, when cavitation starts to form, the acceleration drops to a steady value of 79m/sec^2. The *effective amplitude* of the poker vibrator is therefore significantly reduced by the cavitation effects. This raises the question as to whether the amplitude/frequency ratio of these vibrators can be altered to work at a lower frequency and still be more effective in consolidating the concrete. This is currently being investigated at The University of Liverpool using a pneumatic poker vibrator whose frequency can be altered by reducing the air flow rate.

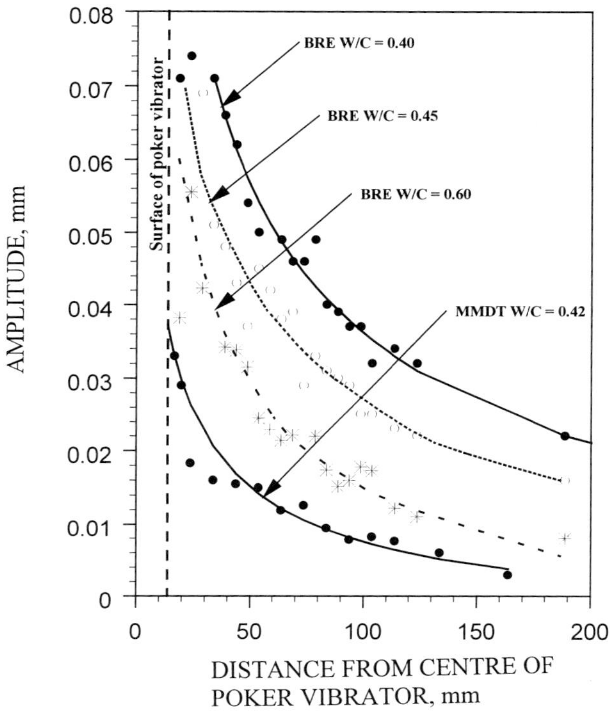

Figure 5 Exponential decay of sinusoidal vibratory waves for BRE and MMDT mixes

The damping coefficients for the BRE mixes decreased from **$13.1x10^{-3}$** to **$4.2x10^{-3}$** as the water-cement ratio decreased from 0.60 to 0.40, see Figure 7. This is somewhat surprising since the two point tests have shown that the plastic viscosity (h) values of the mixes increase with a decrease in the water-cement ratio. The only possible explanation that was offered at the time of the BRE tests was that due to the transmission of mechanical stresses in fresh concrete, according to Bache(8), being influenced by such parameters as elasticity, cohesion, friction, viscosity (shear and bulk), density, damping, and sound velocity. The plastic viscosity (h) value measured by the two point test may gradually decrease at higher oscillating stresses when the bond strength between particles becomes insufficient to prevent flow. In this case the frictional forces between aggregates, especially the sand, may become more dominant than the viscosity of the paste. The frictional forces would then be expected to reduce, since the sand content of the BRE mixes reduces, as the water-cement ratio is reduced. This does not however appear to be true for all the mixes as the effective amplitude is much lower for the superplasticized MMDT mixes than it is for the BRE mixes. The reason for this is not yet clear. Future tests will concentrate on recording more accurate acceleration values at distances less than 20mm from the poker vibrator in order to clarify these uncertainties.

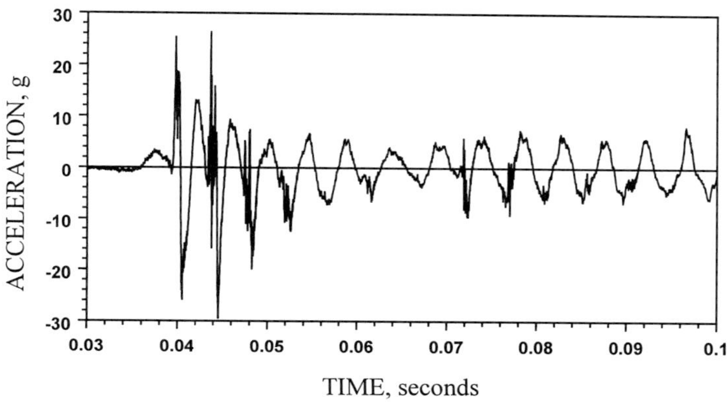

Figure 6 The effective amplitude of the poker vibrator is significantly reduced by cavitation effects

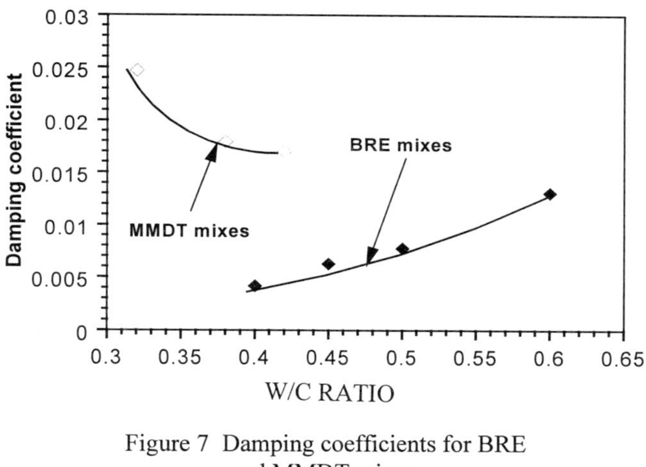

Figure 7 Damping coefficients for BRE and MMDT mixes

The damping coefficients of the superplasticized MMDT mixes are much higher than for non-superplasticized concretes. The damping coefficient for a mix with a water-cement ratio of 0.32 is $\mathbf{24.7x10^{-3}}$. The damping coefficients of the MMDT mixes also increase with decreasing water-cement ratios. This relates well to the increase of the plastic viscosity (h) values of these that were obtained using Tattersall's two point test apparatus. It is not surprising that these show a better relationship with the plastic viscosity (h) values since they all have the same paste content unlike the BRE mixes that the paste content increased as the water-cement ratio was reduced. It may therefore be concluded that there is a relationship between the plastic viscosity (h) values and the damping coefficients of concretes, as long as the paste content is the same for all concrete mixtures.

CONCLUSIONS

The workability obtained by the use of high dosages of superplasticizers is not directly values increased despite the concrete mixtures all having a slump of 150mm. The rate of increase of these values when reducing the water-cement ratio was higher when increasing the dosage of the superplasticizer than when increasing the paste content to retain the same slump.

The damping coefficients of sinusoidal vibratory waves were found to be higher for superplasticized rather than non-superplasticized mixes despite having similar slump values. This has practical implications regarding the consolidation of high strength concrete mixes on site. The radius of influence of the poker vibrator is reduced and therefore the insertion points need to be closer together than normal strength mixes.

The amplitude generated around the vibrating head was found not to be the same as the amplitude of the vibrator. The reduction of the energy transmission from the vibrator to the fresh concrete is believed to be due to cavitation effects occurring at high vibration intensities. Whether the amplitude/frequency ratio of commercial poker vibrators can be altered to work at a lower frequency and still be more effective is currently being investigated at The University of Liverpool.

ACKNOWLEDGEMENTS

Thanks are due to the EPSRC for their financial support for this work (Grant Ref. No. GR/L55223) which is continuing at The Department of Civil Engineering at The University of Liverpool. The authors would also like to thank Dr. G. H. Tattersall for his invaluable advice during the project, and Mr. Jason Lowe and Mr. Eric Phua for their assistance with experimental work described in this paper.

REFERENCES

1. ACI COMMITTEE 309, Standard practice for consolidation of concrete, ACI Manual of Concrete Practice, 1983, pp. 309-1 to 309-40.

2. TEYCHENNE, D C., FRANKLIN, R. E., ERNTROY, H. C., Design of Normal Concrete Mixes, Building Research Establishment, Department of the Environment, 1988, p. 44.

3. SOUTSOS, M N.; AND DOMONE, P L J., Design of high strength concrete mixes with normal weight aggregates, Proceedings of the 3rd International Symposium on the Utilization of High Strength Concrete, Lillehammer, Norway, June 20-23, 1993, Norwegian Concrete Institute, 1993, Vol. 2, pp. 937-944.

4. DOMONE, P L J.; AND SOUTSOS, M N., An approach to the proportioning of high strength concrete mixes, Concrete International, Vol. 6, No. 10, October 1994, pp. 26-31.

5. TATTERSALL, G H. AND BANFILL, P F G., The Rheology of Fresh Concrete, Pitman Press, 1983, p. 356.

6. DESSOFF, M., Sur l'etude de la pervibration du beton, Annales des Ponts et Chaussees (Paris), No. 5, 1937, pp. 681-688.

7. ACI COMMITTEE 309, Behaviour of fresh concrete during vibration, ACI Manual of Concrete Practice, 1983, pp. 309.1R-1 to 309.1.R-18.

8. BACHE, H H., The Processing of Fresh Concrete, Concrete Research Laboratory, Karlstrup, Jan. 1973.

MODELLING AND TESTING OF FLOOR SYSTEMS WITH CEMENTITIOUS SCREEDS AND ADHESIVES

J M J M Bijen

E Schlangen

INTRON B.V.

T Salet

FEMMASSE B.V.

Netherlands

ABSTRACT. Restrained movements of concrete, screeds and top layers lead to stresses. Sometimes these stresses exceed adhesion and tensile strength, causing cracks and delamination. An investigation was conducted to model the performance of a floor system composed of a concrete slab, a plastic film, a cementitious screed, three types of adhesives and a ceramic tile. The modelling was carried out using the computer module MLS 2.5D, which is based on a finite element program. A testing program was performed to generate input data for the computation and to check the predicted performance.

Keywords: Concrete, Adhesives, Screed, Tiles, Floor, Shrinkage, Creep.

Professor Jan Bijen is director of INTRON BV the Dutch institute for quality assessment in the building industry. He is also part-time professor in materials science at the Faculty of Civil Engineering and Geoscience of Delft University of Technology. His field of expertise is the durability and sustainability building materials.

Theo Salet is a partner in FEMMASSE B.V.

Erik Schlangen is senior project engineer in the consultancy and development department of INTRON. He is a specialist in materials engineering.

INTRODUCTION

Multi-layer systems such as concrete floor slabs with screeds and tiles exhibit complex behaviour under practical conditions. The substrate, screed and tiles tend to move, causing stresses in the components of the floor systems, which can sometimes lead to failures. Guidelines for applying flooring systems are often based on long-term experience and are of an empirical nature. When the systems are changed, these empirical rules may cease to apply.

INTRON and its affiliate FEMMASSE BV have developed a computer module Multi-Layer System (MLS), which computes the strains and stresses within and between layers.

At the request of Schönox Nobel GmbH of Germany this program was used to model the performance of a floor system composed of a concrete slab, a plastic foil, a cementitious screed, three types of adhesives and ceramic tiles [1].

A number of properties of adhesives and screed were tested to generate input for the computations as well as to check the predicted performance experimentally.

FLOOR COMPOSITION AND MATERIALS

System

Figure 1 shows a cross section of the flooring system applied. It is an unbonded floor due to the presence of the plastic film, which reduces shear stresses and moisture transport.

Figure 1 Schematic cross-section of floor system

Adhesives

The following types of adhesives were tested experimentally:

- powder adhesive SK
- powder adhesive PCI
- Plasticol 14

The adhesives were applied with a comb to a thickness of 2.0, 1.4 and 1.8 mm respectively.

Screed

The screed was composed of cement (CEM I 32.5R, 0.4 kg), sand and gravel up to a maximum grade size of 8 mm (1.6 kg). The Young's modulus was assumed to be 30 kN/mm² with a Poisson contraction coefficient of 0.2. A screed thickness of 40 mm was assumed.

Tiles

Large-size tiles of 8 mm thickness were assumed for most of the numerical analyses. In fact, this was a situation where there were no joints. An analysis of the effects of the presence of joints was made, but will not be discussed here. The Young's modulus of the tiles was assumed to be 15 kN/mm² with a Poisson contraction coefficient of 0.2.

Specimens

The specimens were manufactured in accordance with the guidelines of Schönox Nobel GmbH.

SHEAR AND TENSILE TEST

Direct Tension Test

The tests were performed under load-controlled conditions, with a continuously increasing load. The ultimate strength at failure at ages of 1, 3, 6 and 28 days is shown in Table 1.

Table 1 Ultimate strength (mean of 5) of the adhesives loaded in a direct test at various ages

AGE AT LOADING	TYPE OF ADHESIVE		
	SK (N/mm²)	PCI (N/mm²)	Plasticol 14 (N/mm²)
1 day	0.28	0.28	0.15
3 days	1.02	0.38	0.35
6 days	2.03	0.75	0.48
28 days	2.40	1.50	1.11

Note that the type of failure differs for each type of adhesive. The SK type of adhesive actually does not fail. It is stronger than the screed. Failure occurs in the cementitious screed. The PCI type of adhesive commonly fails in the adhesive itself, whilst the Plasticol 14 fails in the interface with the screed. Obviously, the adhesion between Plasticol 14 and the cementitious screed is not sufficient to benefit from the full strength of the adhesive. It was noted that this might be caused by the method used to apply the adhesive.

Shear Test

The shear strength-strain relationship was tested by loading the samples in torsion. A schematic view of the specimen and the testing rig is shown in Figure 2. The tests were deformation-controlled, with a continuously increasing displacement. The tests were ended arbitrarily some time after the ultimate load was reached. Three samples were loaded in torsion at the ages of 1, 3, 7 and 28 days.

Figure 3 shows typical shear stress-strain relationships for the three adhesives at the age of 28 days. The ultimate shear strength, shear strain and shear modulus (G-modulus) are summarised in Table 2.

The G-modulus is defined as the tangent of the stress-strain curve between 10% and 90% of the failure load. The ultimate shear stress is computed from the torsion by means of the following equation:

$$\tau_u = (M_w \cdot r)/I_p,$$

with :
- τ_u : Ultimate shear stress
- M_w : Torsion moment
- r : Radius of the samples (25 mm)
- I_p : Polar moment of inertia being $1/32*\pi*d^4$
- d : Diameter of sample (50 mm)

Figure 2 Photograph of the specimen and testing rig used to load samples in torsion

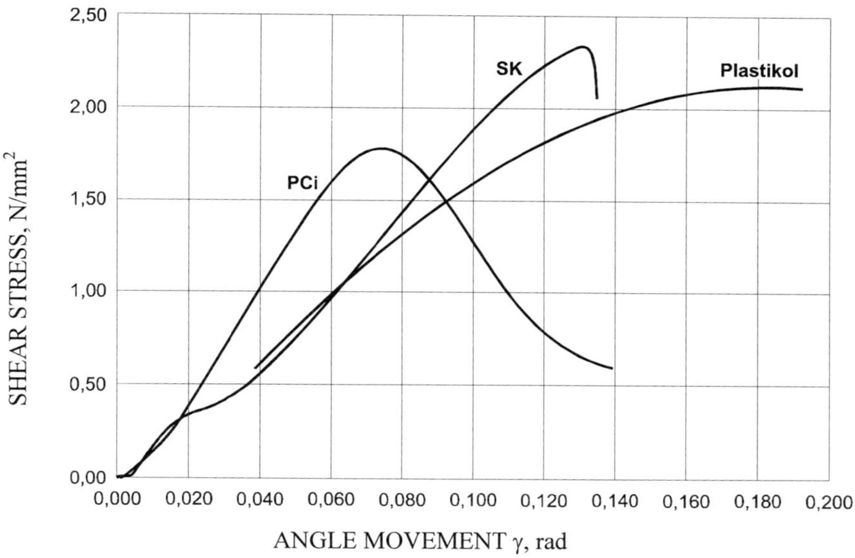

Figure 3 Shear stress as a function of angle displacement

Table 2 Mean values of ultimate shear strength (τ_u), ultimate strain (γ) and shear modulus (G) of the adhesives loaded in torsion

AGE	TYPE OF ADHESIVE								
	SK			PCI			Plasticol 14		
	τ_u (N/mm²	γ (rad)	G (N/mm²)	τ_u (N/mm²)	γ (rad)	G (N/mm²)	τ_u (N/mm²)	γ (rad)	G (N/mm²)
1 day	0.29	0.036	9.8	0.10	0.163	0.75	0.09	0.11	0.54
3 days	1.64	0.064	35.6	0.36			0.16	0.134	1.27
7 days	2.87	0.066	45.9	0.35	0.085		0.35	0.183	6.57
28 days	2.90	0.072	42.8	1.6	0.071	26.1	1.7	0.165	11.4

Note
The tests have shown that Plasticol 14 has considerable after-peak strength. Therefore it will also have a fairly large additional failure energy that has been neglected in this study! The results of the numerical analyses are conservative, particularly in the case of Plasticol 14.

HYGRAL SHRINKAGE OF THE SCREED

The hygral shrinkage of the screed exposed in a climate room to 20^0C and 50% RH after one day at 20^0C in the mould was measured over the full length of 40x40x160 mm bars. The average result is shown in Figure 4.

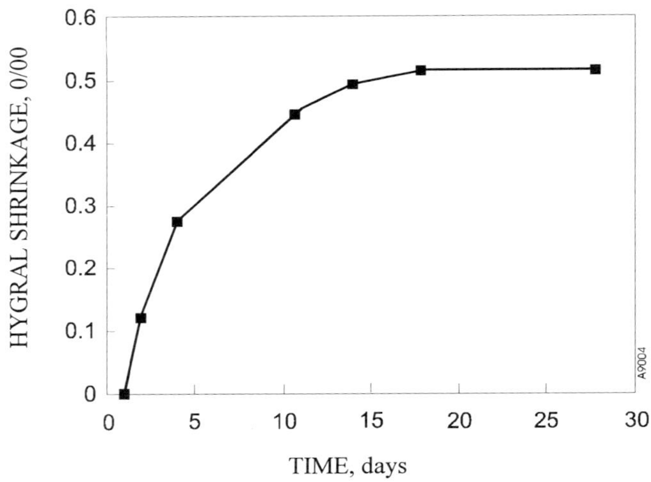

Figure 4 Mean drying shrinkage of the cementitious screed

NUMERICAL ANALYSES

Module MLS/2.5D of the finite element program FEMMASSE was used for the numerical analyses. This program is the property of FEMMASSE B.V. MLS was developed by Roelfstra; background information can be found in references 2 to 4.

The tiles were modelled as a continuous layer connected to the screed by means of an interface as shown in Figure 5. The interface properties are matched with the data from the experiments described in the previous section. Details of the so-called cohesion slip model of the interface can be found in the manual of the module MLS/2.5D [4].

Quasi-instantaneous analyses were performed in which a gradually increasing volume shrinkage was ascribed to the cementitious screed only (1 mm/m per 6 'hours').

Strain and stresses were computed. Examples of shear stresses and normal stresses computed are presented in Figures 6 and 7.

Typical data from the test results have been assembled in Table 3.

The symbols that are used in this table are explained in Figure 8.

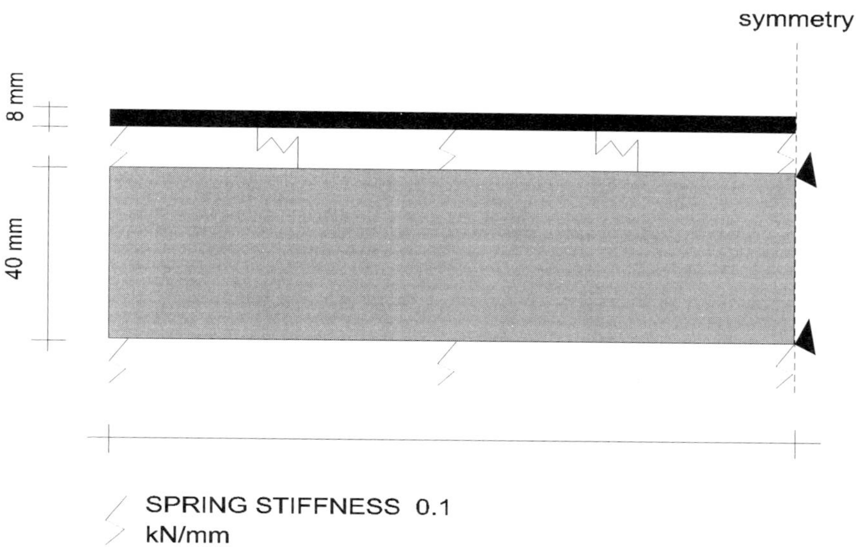

Figure 5 Schematic view of the numerical model

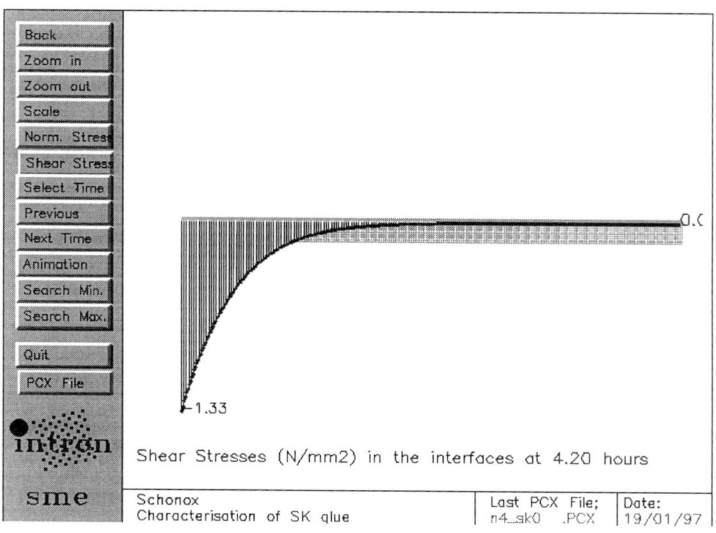

Figure 6 State of the ultimate shear and normal stresses in the interface at the failure load in the case of adhesive SK

Table 3 Summary of numerical results of ultimate states

PARAMETER			TYPE OF ADHESIVE			
			SK	CPI	Plasticol 14	Plasticol 14 *
Adhesive	a	(mm)	340	390	560	670
	τ	(N/mm²)	-1.30	-0.86	-0.71	-0.84 (0.96**)
	σ	(N/mm²)	-0.43	-0.28	-0.19	-0.22
Tile	σ (max) (N/mm²)		-17.2	-12.3	-15.8	-23.2
Screed	σ (max) (N/mm²)		4.5	3.2	4.1	6.0
Differential shrinkage at failure (mm/m)			0.7	0.5	0.65	0.95

Note 1:a is arbitrarily defined at the length needed to transfer 98 % of the shear force.
Note 2: * means based on shear strength that has been measured instead of tensile strength
Note 3: ** means that a maximum value of 0.96 was found prior to softening

Note that the differential shrinkage, as given in Table 3, is not yet a true design value. Among other things, a long-term strength must be considered and a certain safety margin allowed for failure.

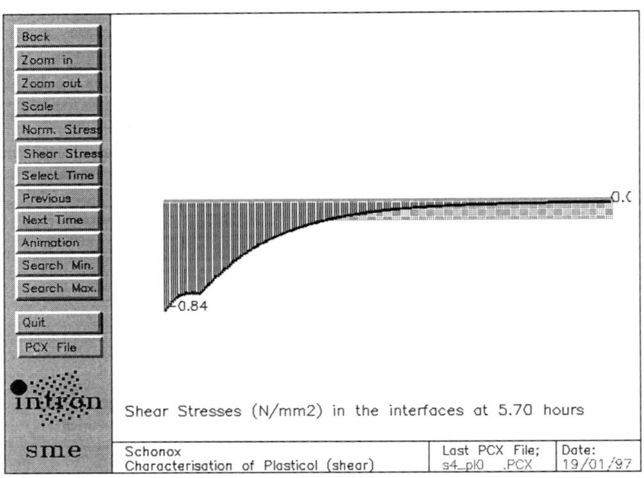

Figure 7 State of the shear and normal stresses in the interface at the failure load in the case of adhesive Plasticol 14. Failure state based on shear strength (1.7 N/mm²)

The numerical analyses have shown that Plasticol 14 is indeed capable of withstanding the largest differential shrinkage of the adhesives that have been tested, where a proper bond has been guaranteed. In a previous section a shrinkage of 0.5 mm/m was measured for the cementitious screed. It should be noted from Table 3 that a differential shrinkage of this kind can be transferred by all the adhesives tested. In principle, the tiles can be glued immediately after the screed has hardened sufficiently, but not much of a safety margin seems to be left in the case of the CPI type of adhesive .

In practice, shrinkage values of up to 1 mm/m cannot be ruled out. In the event of such high shrinkage values, a certain drying period is needed before applying adhesives and tiles in order to limit the differential strain.

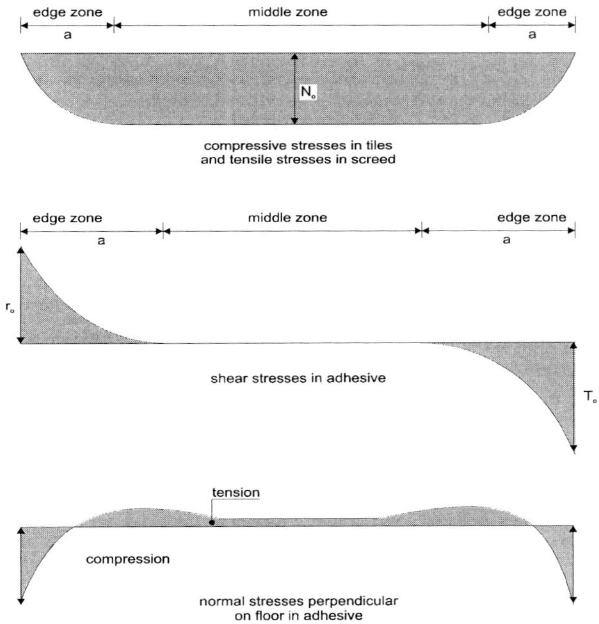

Figure 8 Definitions of symbols used in Table 3

ADVANCED NUMERICAL ANALYSES

The results presented in the previous sections show that Plasticol 14 will be capable of withstanding the largest differential shrinkage, where a proper bond is guaranteed. However, it has also been shown that the tensile stresses in the cementitious screed seem to exceed its tensile strength. Realising that the numerical analysis discussed is based on instantaneous behaviour only and does not take into account time-dependent shrinkage, creep and relaxation it was decided to perform advanced numerical analyses taking creep (relaxation) into account. The effects of its presence were also analysed but not subsequently discussed. Reference is made to [1].

Time-dependent analyses were performed, taking the actual shrinkage rate and creep (relaxation) into account. The mean evolution of the moisture potential, taken over the full thickness of the screed, is shown in Figure 9.

Since large-size tiles were assumed, drying was computed in a horizontal cross-section only. This is an acceptable approximation since the size of the tiles is much larger than the thickness of the screed.

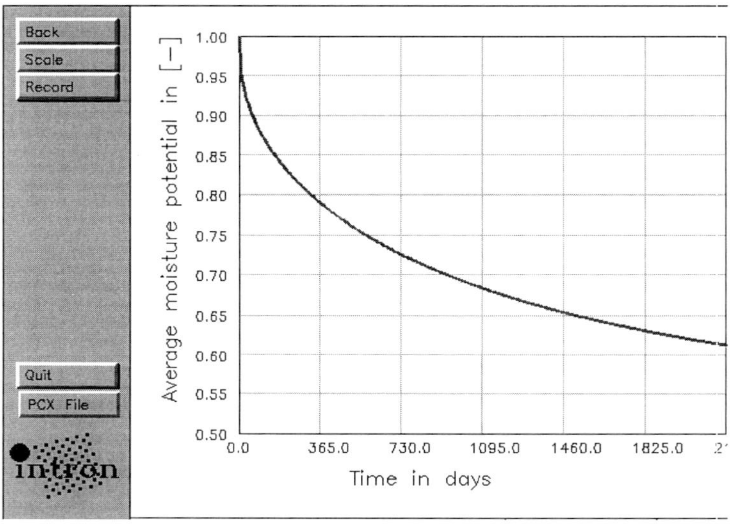

Figure 9 Mean evolution of the moisture potential in the screed, computed from a 2D analysis

The mean evolution of the moisture potential was ascribed to the screed in order to make stress analyses. Creep of the screed was taken into account by means of an ageing Maxwell chain. The creep data were approximated, based on the CEB-FIP model code 1990 for concrete (fc=10 N/mm²). Relaxation was computed as e^{-}.

The stresses in the screed and tiles, as well as the shear and normal stresses in the adhesive, are plotted in Figures 10 and 11, respectively. Note that the analysis was performed for the adhesive Plasticol 14 only.

When compared with the results from the previous sections, it is obvious that the creep of the screed reduces considerably the stresses in the adhesive, screed and tiles.

A maximum shear stress of 0.28 N/mm² was computed after 6 years, which means that the screed is most unlikely to crack since the tensile strength is in the order of magnitude of 1.7 N/mm².

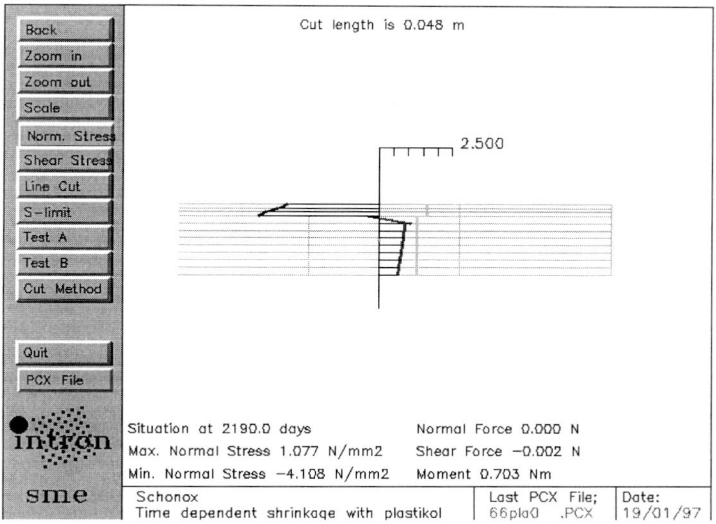

Figure 10 Stresses in tiles and screed, at a certain distance from the free end, beyond the transfer zone. Note that shrinkage is 0.4*0.0015 mm/m = 0.6 mm/m after 6 years

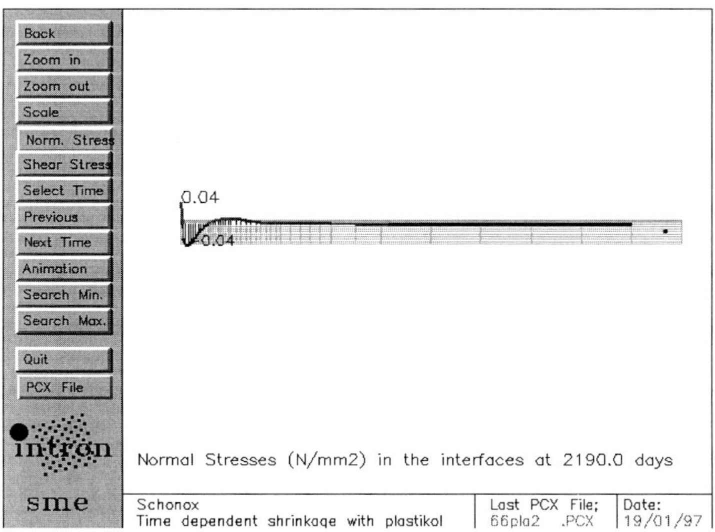

Figure 11 Shear and normal stresses in the adhesive (Plasticol 14).Note that shrinkage is 0.4*0.0015 mm/m = 0.6 mm/m after 6 years

CONCLUSIONS

Plasticol 14 is expected to be the most suitable type of adhesive among the types of adhesive tested for withstanding a large differential shrinkage between a cementitious screed and tiles, provided a proper bond with the cementitious screed is guaranteed.

In the literature the maximum differential shrinkage acceptable for the adhesives is commonly based on a simple instantaneous analysis. Where time-dependent behaviour is not considered, it has been shown that the maximum differential shrinkage of the various types of adhesives is:

- SK 0.7 mm/m
- PCI 0.5 mm/m
- Plasticol 140.65 mm/m (poor bond) and 0.95 mm/m (proper bond)

However, instantaneous analyses also show large tensile stresses in the cementitious screed and even high compressive stresses in the tiles. On the basis of instantaneous analyses, cracks are to be expected.

Where the creep of the screed, time-dependent drying shrinkage and the stiffness of the joints are taken into account, it appears that the principal stresses in the screed, tiles and adhesives are considerably reduced and no cracks are expected.

In general, this type of modelling contributes to an understanding of the major factors affecting the performance of floor systems.

ACKNOWLEDGMENT

The authors would like to express their gratitude to Dr Meinert and Dr Hilgenbrink of Schönox Nobel GmbH, who gave permission to publish these results and to Dr Peter Roelfstra of FEMMASSE BV for his assistance.

REFERENCES

1. Performance of a floor system composed of a cementitious screed, tiles and different types of adhesives, INTRON Report No. 97015, 1997.

2. ROELFSTRA, P E, A numerical approach to investigate the properties of concrete, thesis presented at University of Technology of Lausanne, No 788, 1989.

3. ROELFSTRA, P E, WITTMANN, F H, Some aspects of the concept of total fracture energy, Vorträge der 17. Sitzung des Arbeitskreises Bruchvorgang, DVM-SVTM, Basel 1998, pp. 345-356

4. Manual MLS/2.5D of the finite element program FEMMASSE, a proprietary program of FEMMASSE BV, The Netherlands.

THEME TWO:

NEW PRACTICES

TEST-METHODS ON ROAD CONCRETE WITH A VERY POROUS STRUCTURE

E De Winne

Ghent University

Belgium

ABSTRACT. Research is carried out in the Magnel Laboratory – Ghent University into the properties of road concrete with a very porous structure, enriched with polymer emulsions. The aim of the tests is to optimise the composition of the porous concrete. In this context, the choice of the type of polymer emulsion is the most important question. The criteria for this choice are mainly the mechanical properties, setting behaviour and durability. The research had six broad phases: exploratory tests, evaluation of the polymer emulsion, tests on simulated road surfaces, durability tests, establishing additional properties and definitive composition. In 1996, six different wearing courses were cast over a length of a 3 km long test-section in Flanders (Belgium). Four different wearing courses, each 4 cm thick, were cast on a continuously reinforced concrete pavement. In addition to the conventional wearing courses in Stone Mastic Asphalt 0/14 and Drainage Asphalt 0/14, a VERY POROUS CONCRETE 0/7 and a FINE CONCRETE 0/7 were cast, wet on wet, on the unset continuously reinforced 18 cm base course. The composition and the casting techniques of these concrete mixtures are innovative. The zone, containing four test sections, is bordered by 2 test sections of bituminous composition with a wearing course: STONE MASTIC ASPHALT 0/7 and DENSE ASPHALT 0/10 with crushed gravel chippings 10/14. The comparison results show that it is possible to construct pavements in cement concrete which are as noiseless as bituminous pavements. It has been proven that porous cement concrete pavements, cast wet on wet, are feasible from a technical point of view. All test results, the course of the contraction, the test results about the resistance to traffic damage, and also the noise measurements, will be given.

Keywords: Concrete, Porous concrete, Asphalt, Porous asphalt, Road pavements.

Professor Etienne De Winne is Head of the Road Department, Faculty of Applied Sciences, Ghent University, Belgium and General Inspector of the Ministry of the Flemish Community, Environment and Infrastructure Department, Road Infrastructure and Traffic Administration, Brussels, Belgium. He specialises in the testing, durability and protection of roads.

INTRODUCTION

Laboratory Tests

Research is carried out in the Magnel Laboratory – Ghent University into the properties of road concrete with a very porous structure, enriched with polymer emulsions. The aim of the tests is to optimise the composition of the porous concrete. In this context the choice of the type of polymer emulsion is the most important question. The criteria for this choice are mainly the mechanical properties, setting behaviour and durability. In the first phase, by way of introduction, a few test mixtures have been made up on the basis of a mix-type known from the literature, with a view to identifying their mechanical properties and processing behaviour.

In the second phase, after confirmation of the basic suitability of such a mixture for use, more extensive series of tests have been carried out with four polymer emulsions for comparison of mechanical properties. In the third phase, these mixtures have been used in simulated road surfaces of restricted dimensions, mainly with the aim of ascertaining how they interact with the substrate. In the fourth phase, durability tests have been carried out on cores bored out of the simulation test samples. The fifth phase consisted of the establishing of additional properties of the very porous concrete mixtures used in the previous phases. In the sixth phase, some of the performance tests and durability tests have been repeated on the definitive mixture chosen, with a modified composition.

Test Sections

The Infrastructure Department seized the opportunity to test cement concrete pavements and to compare its acoustic, road constructional and financial aspects with those of bituminous pavements, by constructing test sections on a road with a single lane of 3 m in each direction (5000 vehicles a day with 12% heavy freight traffic).

Four different upper layers, each 4 cm thick, were cast upon a continuously reinforced cement concrete pavement with a thickness of 18 cm. Each of those 4 test sections had a width of 2 x 3 m:

1. On the first test section, 834 m long, a very porous asphalt concrete layer 0/14 is cast on the continuously reinforced cement concrete pavement with intermediate SAMI (Stress Absorbing Membrane Interlayer), which meanwhile has become a very classical composite pavement.
2. On the second test section, 286 m long, a stone mastic asphalt layer 0/14 is cast on the continuous reinforced cement concrete pavement.
3. On the third test section, 832 m long, an upper layer in fine concrete 0/7 is cast, wet on wet, this being a technique which was already applied in Austria in 1990. The surface treatment of the upper layer is carried out according to the "washing-out" method, which was improved in Belgium.
4. On the fourth test section, 537 m long, and equally wet on wet, a very porous concrete layer 0/7 is applied. Such an upper layer was already studied in the Netherlands from 1987 on, and constituted the subject of an European study project - the BRITE-EURAM project BE 3415 - in which the Netherlands as well as Germany and Spain have co-operated.

The 4 sections are delimited by test sections with a bituminous casting, yet with different bituminous upper layers:

5. On one side, a test section, 269 m long, with an upper layer in dense asphalt 0/10 with chippings 10/14.
6. On the other side, a test section, 407 m long, with an upper layer in stone mastic asphalt 0/10.

TENDER SPECIFICATIONS

1. Major principles concerning the under layer in continuously reinforced concrete with a nominal thickness of 18 cm (used in all test sections)

- cement: CEM I 42,5 LA or CEM III/A 42,5 LA yet of the same nature as the one used in the top layer in very porous concrete or fine concrete. Quantity: 375 to 425 kg/m³ of concrete.
- chippings: porphyry; maximum grain size 32 mm.
- air entraining agent: the addition is compulsory; the enclosed air content is 3,5 to 6%, with a distance factor smaller than 0,2 mm.
- longitudinal reinforcement: diameter 18 mm with a 15 cm intermediate distance, the upper part of the longitudinal bar being 12 cm above the foundation surface.
- transversal reinforcement: diameter 12 mm with a 70 cm intermediate distance.
- anchorage joints: 4 joints each 0,65 m wide and 0,90 m high at each end.
- compression strength : the characteristic compression strength amounts up to 52 N/mm²; the required average compression strength amounts up to a minimum of (52 + 1,645.s) N/mm² (in each case after 90 days and on cores ∅ 100 cm²).

2. Major principles concerning the upper layer in very porous concrete

- cement: CEM I 42,5 LA or CEM III/A 42,5 LA. Quantity: 300 to 350 kg/m3 of concrete.
- porosity (accessible hollow spaces): minimum 25% - maximum 30%.
- chippings: porphyry 4/7 or 7/10.
- grading: the grading of the aggregates shows signs of discontinuity.
- polymer emulsion: polymer mass: 10 to 12% of the cement mass.
- compression strength: 18 N/mm² (after 28 days and on cubes of 158 mm).
- flexural tensile strength: 4 N/mm² (after 28 days and on prisms of 100x100x400 mm).
- splitting strength: 2 N/mm² (after 28 days and on cores of 100 cm²).
- dynamic modules of elasticity: 20 to 25 kN/mm².
- compaction index: 1,20.
- processing: within the half hour after the mixing.
- protection against dehydratation: waterproof foil during at least 72 hours.
- levelness: with LPA: VC (10m) ≤ 70
VC (2,5m) ≤ 35
- roughness: with odoliograph DWC ≥ 0,45 at 50 km/h.

3. Major principles concerning the upper layer in fine concrete

- cement: CEM I 42,5 LA or CEM III/A 42,5 LA. Quantity: 425 to 450 kg/m³ of concrete.
- chippings: porphyry 4/7 - a minimum of 55% of the sand - chippings mixture.
- sand: fine, round grains, in an as low as possible percentage, however without hampering its good processing.
- air content: 5%, with a distance factor smaller than 0,2 mm.
- compression strength: 50 N/mm² (after 28 days on cubes of 158 mm)
- flexural tensile strength: 5,5 N/mm² (after 28 days and on prisms of 100x100x400 mm).
- supersmoother: longitudinal wiper beam compulsory.
- surface treatment: washing out, during which an average levelling depth of 1,2 mm is aimed at.
- levelness: with LPA: VC (10m) ≤ 70
VC (2,5m) ≤ 35
- roughness: with odoliograph: DWC $\geq 0{,}45$ at 50 km/h

CONCLUSIONS

Laboratory Tests

Phase 1: Exploratory tests

The aim of the first exploratory tests was to verify that the assumed composition for the porous concrete produces a serviceable mortar. The conclusion was that, visually, the test sample has a homogeneous appearance and substantial cohesion. On the basis of these observations it is concluded that the basic composition put forward is usable.

Phase 2: Evaluation of the polymer emulsion

The aim of this part of the research was to compare some elementary properties of the porous concrete produced with the four emulsions. Before final mixtures and test samples were prepared, compaction methods have been experimented. The aim of the compaction was not to maximise compression strength but to obtain a sufficiently firm and cohesive concrete with a suitably porous structure. So compaction on the vibrating table is chosen.

Phase 3: Tests on simulated road surfaces

With three polymers (acrylic emulsion, styrene-butadiene emulsion, styrene-acrylic emulsion) six non-reinforced test slabs have been made up in the laboratory, and with these slabs the adhesion behaviour of the porous concrete on the substrate is tested.

The test results (compression strength, mass/volume) were collected at 7 and 28 days old for the different polymers.

Phase 4: Durability tests

The remaining four series of three test cylinders for each polymer are intended for four different durability tests:

- a test to establish resistance to road salt; these test results were expressed as loss of mass per unit of upper surface exposed to the salt solution.
- a frost-resistance test by alternate freezing and thawing; the results of the visual observations were summarised and photos show halves of the test samples after they have been sawn through.
- a test to establish resistance to changing temperatures (tensile strength).
- a wet-dry test (tensile strength).

Phase 5: Establishing additional properties

To complete the knowledge of the basic properties of the material, a few additional tests have been carried out on mixtures with the three used polymers.

- tests after storage in the laboratory atmosphere (compression tests, air-dry mass/volume, bending tests, splitting test);
- tests after storage under water (bending tests, compression tests);
- establishing (degree of) contraction (basic measurements, contraction factor).

Phase 6: Definitive composition

After an evaluation of the available results, it is proposed to raise the polymer (solid material) content from 10 to 12 %.

- porphyry crushed stone 4/7: 1500 kg/m³
- sand 0/1: 100 kg/m³
- cement CEM III A/42.5: 300 kg/m³
- water: 60 kg/m³
- polymer emulsion: 72 kg/m³

It is also proposed to use styrene-acrylic emulsion polymer.

With this composition, most of the following tests have been repeated:

- the production of a simulated road surface (1 m²) with a waiting time of two hours;
- compression tests on cubes compacted:
    * with the stamper;
    * with the vibrating table and loading with a steel weight;
- establishing the dry mass/volume on three cubes compacted with the vibrating table method;
- bending, compression and splitting tests on prisms and cubes;
- bending and compression tests at twenty-eight days on and cubes, stored in water;
- tensile tests at twenty-eight days on bored-out cylinders;

- durability tests:
  - road-salt resistance;
  - resistance to frost;
  - resistance to temperature changes with adhesion tests;
  - wet-dry test with adhesion tests.

Test Sections

Conclusions with regard to the road constructional aspects.

The continuously reinforced concrete pavements acting as a base course of the 4 different upper layers, has been cast with conventional slipform paver over a width of 3,25m.

The fine concrete upper layer was cast wet on wet on the base course, in other words immediately following on the casting of the base course, equally with a slipform paver with vibrating heads yet over a width of 3 m. The outer edge of 0,25 m is used as a water drain. The works were carried out without any appreciable problem.

The very porous concrete wearing course was equally cast wet on wet with a slipform paver with a vibrating beam over a width of 3 m. The length of time between the casting of the continuously reinforced concrete and of the very porous concrete had to be adapted in function of setting of the concrete underneath: in other words strongly depending on weather conditions. A maximum of a 4 hours 15 minutes time difference was registered, where as a minimum of 2 hours appeared to be necessary to avoid the mixing of the base and wearing courses. It has to be noted that the temperature during the casting remained under 15°C which is an advantage when casting very porous concrete; this kind of concrete is very heat-sensitive due to the polymer present in the concrete mixture.

Laying a plastic sheet on the base course during this interval might lessen the dependence of weather conditions (such as rain). Anyway, the danger subsists that for any reason (such as a mechanical or electric failure of the slipform paver) the casting of very porous concrete can not be carried out wet on wet, which would result in a hardened unfinished zone of at least 2000 m base course.

Fortunately, the works on this test section were not hampered by any appreciable problem.

Conclusions regarding acoustical aspects

The average arithmetical value of the rolling noise, measured according to the trailer method over the total length of each test section and at 4 different testing speeds, is represented in the subjoined histogram. (Figure. 1)

The rolling noise measured according to the SPB (Statistical Pass By) method has been represented as an example in the histogram with values calculated at 70 km/h, most of the observations were carried out at this speed. (Figure 2)

The regression lines found according to the SPB (Statistical Pass By) method of the maximum rolling noise versus speed logarithm are clustered for each test section and each vehicle category. (Figure 3)

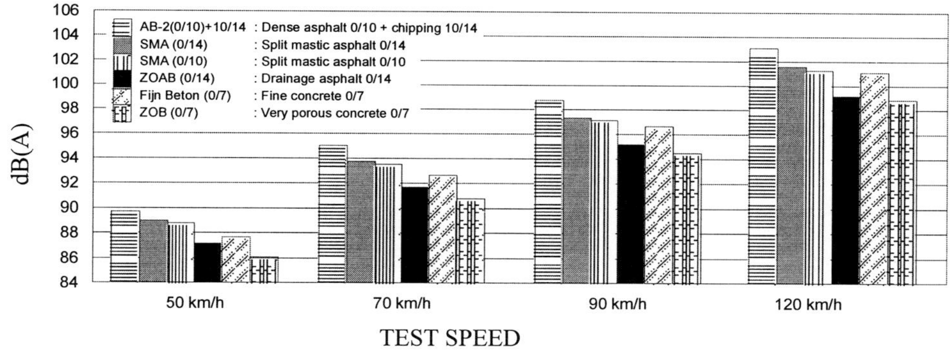

Figure 1 Rolling noise according to the trailer-method 16.9.1996

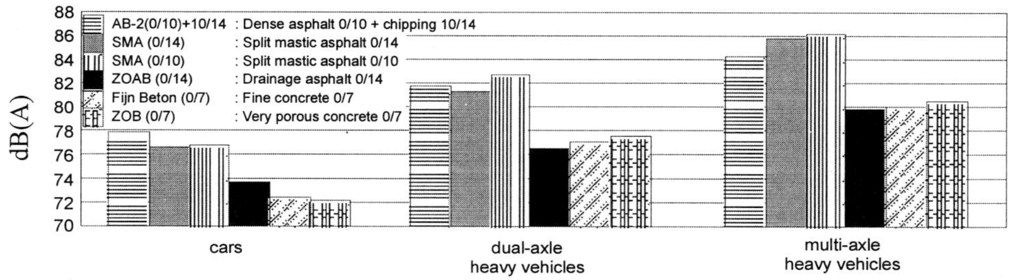

Figure 2 Rolling noise according to the SPB-method 1996 at 70 km/h

The spectral composition of the rolling noise measured according to the SPB (Statistical Pass By) method (yet with the microphone installed at a height of 5 m) is represented in a separate diagram. The results are based upon measurements at 80 km/h with the exception of the test section in dense asphalt where they were based upon measurements at 50 km/h, adjusted as the speed increases up to 80 km/h. (Figure 4)

The results of the test sections split mastic asphalt 0/10 and split mastic asphalt 0/14 show no appreciable difference. The porous pavements (both very porous cement concrete and drainage asphalt) show large variations over the longitudinal section, the extend of the absorption maximum of both porous pavements being rather low; the noise reducing effect, especially with drainage asphalt, shows only at high frequencies (>1250 Hz) and too little at low frequencies (<1000 Hz), precise level at which the optimum noise absorbing drainage asphalt should compensate the more important part of texture induced vibration.

The fine concrete pavements offers positive acoustical results, not only in relation to more coarse concrete pavements but also in relation to bituminous pavements. The constructed cement concrete pavements can certainly be qualified as noiseless pavements and can be compared with noiseless bituminous pavements

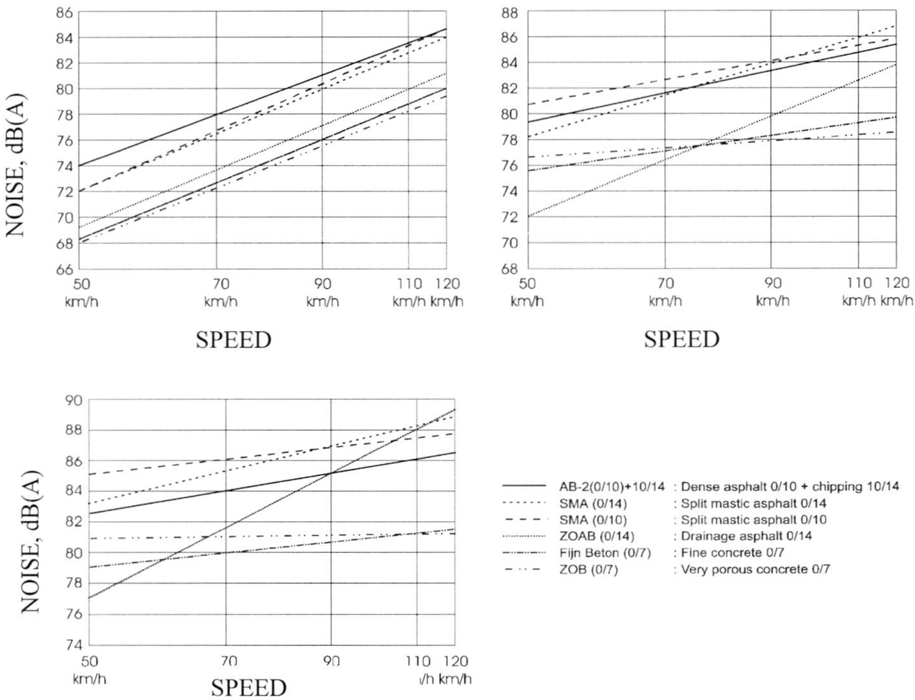

Figure 3 Rolling noise according to SPB-Method (a) cars, (b) dual-axle heavy vehicles, (c) multi-axle heavy vehicles

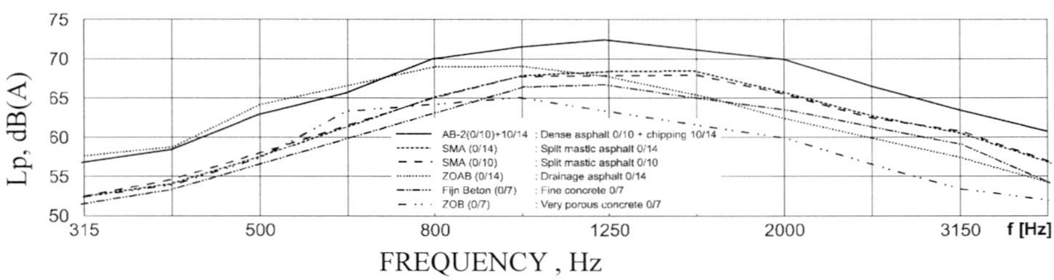

Figure 4 Spectrums according to the SPB-method 27.11.1996

Conclusions regarding the financial aspects

Certain road constructional limiting conditions obviously have an indirect influence on the cost of the pavements. The location of the concrete mixing plant, which has to be as close as possible of the construction site (processing within ½ hour), will play a part in the price determination of a very porous concrete pavement. These hard to define limiting conditions are left out of consideration when comparing prices which only leaves the material and constructing costs to be taken in account. A 22 cm thick fictitious continuously reinforced cement concrete pavement is regarded as a reference pavement.

The construction cost of a very porous cement concrete wearing course with a thickness of 4 cm on a continuously reinforced concrete pavement with a thickness of 18 cm can be estimated at 1,4 times the cost of the reference pavement. Adding polymer to the very porous concrete mixture means a 25 % cost increase.

The construction cost of a fine concrete wearing course with a thickness of 4 cm on a continuously reinforced concrete pavement with a thickness of 18 cm can be estimated at 1,1 times the cost of the reference pavement.

Very porous cement concrete and fine concrete pavements will make a notable contribution to a quieter environment and tests have proved that both achieve a significant noise reduction

PREDICTION OF WEATHERING EFFECTS ON CONCRETE BUILDINGS

E Balodimou

R Ogden

Oxford Brookes University

United Kingdom

ABSTRACT. This paper is based upon work undertaken at the Post Graduate Research School, School of Architecture, Oxford Brookes University. The aim of the work is to develop the basis of a computer model using finite element techniques for the prediction of weathering on concrete surfaces.

The underlying hypothesis is that weathering patterns can be predicted by calculating the frequency and extent of wetting due to rainwater run-off. The amount and speed of rain impinging on a surface, the geometry of a surface and its absorption capacity are critical factors that determine rainwater run-off, and this in turn influences the manner in which particulates are distributed on surfaces.

Keywords: Weathering, Soiling, Prediction, Concrete, Rain run-off, Absorption, Finite element techniques.

Dr Efcharis Balodimou (Dip. Arch) is a researcher specialising on weathering of buildings and its prediction. Her research has focused mainly on concrete buildings.

Dr Raymond Ogden (BaHons, Dip Arch, MCSD) is a Principal Lecturer / Reader in Architecture and specialises in construction, building physics and production engineering. He has been responsible for numerous research projects particularly in the area of steel construction funded by ECSD, DoE, DTI, EPSRC and industry and has published widely.

Both authors are members of the Technology Group of Oxford Centre for Sustainable Development (OCSD).

INTRODUCTION

The alteration of a building's appearance as a result of exposure to atmospheric and environmental conditions is termed *weathering*. These conditions induce visual and compositional changes within and on the surface of building materials that can be of physical, chemical or biological nature and may have a favourable or unfavourable impact upon the appearance and physical state of a building. Most of the factors which induce weathering, i.e. climate, environment, adjacent materials, may be taken into consideration during the design process and can, to some extent, be anticipated, controlled and used as positive elements in design (1-3).

This paper deals with the visual changes on building facades due to the effects of water and accumulation of particulates. These changes are referred to as *soiling.* Soiling can physically damage certain types of surface, but usually there is only a visual change which can nevertheless have an adverse effect on the aesthetic and commercial value of a building (4). It is important to consider these changes during the design process so that the effects of weathering can be optimised. This work proposes a model for prediction of weathering which could help designers optimise their designs with regards to weathering and assumes a wet climate.

THE WEATHERING OF CONCRETE BUILDINGS

Concrete buildings are strongly associated with 'uneven' weathering, which becomes noticeable over a relatively short time, i.e. less than ten years in a wet climate. This can be due to bad detailing and design but is also due to the material's properties and its reaction to the environment. The majority of concrete's weathering problems are related to the presence of water which reacts with the constituents of the material and also redistributes any impurities present on its surface, creating tide marks and stains.

Rainwater runoff is usually the main cause of uneven of soiling of facades. When driving rain impinges on a building facade, some is absorbed into the building material until saturation and the rest forms run-off. Typically this runoff acts as a cleaning agent for the upper parts of buildings and acts as a carrier of particles from the higher regions to those beneath. At the point where run-off stops, soiling stains are created.

Each building material has a different absorption capacity which is expressed by a *water absorption coefficient (A)* defined by C.I.B (5). This absorption coefficient is an important factor which influences the extent to which the building facade becomes soiled, since it determines if and how soon rain runoff occurs, and how far down the facade it will reach. The less porous a building material is (low water absorption coefficient), the faster rain runoff occurs and vice versa.

Absorption coefficients vary between different materials from 0 (zero), i.e. glass and metal, to 300-600 gr/m $sec^{0.5}$, i.e. ordinary brick. When the value of the absorption coefficient is under 10 gr/m $sec^{0.5}$, rain runoff commonly reaches all the way to the ground level, thus cleaning the whole surface of the building facade. Concrete, however, has an absorption coefficient varying from 10 to 40 gr/m $sec^{0.5}$, which means that it has a tendency to create runoff relatively quickly. However, where the amount of rainfall is greater on the top of the facade than on the lower region, runoff may not reach the ground level. Therefore the surface is not evenly cleaned and unsightly stains are formed where the rain runoff stops (6, 7).

A Predictive Model for Weathering

The model for the prediction of weathering proposed in this research may serve as an analytical tool aiming to optimise the weathering performance of buildings. It could be further used in conjunction with existing modelling and rendering packages to visualise weathering effects.

The proposed model attempts to predict the wetting and soiling patterns created on buildings' surfaces. Since most weathering patterns tend to be due to the effects of water - soiling patterns due to water saturation stains and distribution of dirt particles by rain, organic growths usually appearing on parts of a facade which get wet and do not become exposed to enough heat to dry out (usually shaded areas), this strongly suggests that wetting and soiling are the first stages of the weathering process and indicators of possible further decay and deterioration. If areas likely to be wetted and soiled are identified on a facade, the manner in which it will weather could be predicted. Especially in the case of concrete, which greatly reacts with water, it is very important to identify areas most exposed to water, so that unsightly changes can be avoided.

It is suggested that wetting and soiling patterns can be predicted by calculating run-off created on a building surface. A fully working model that would predict wetting and soiling patterns, to be of greatest practical application, should be able to fulfil certain tasks. It should recognise the effects different materials have upon weathering, as well as the influence of a building's geometry. It should also be able to calculate amounts of driving rain impinging on a building when information provided consists of the amount of rain falling in the area and the dimensions of the building. It should be able to calculate the rain distribution upon the facade when the geometrical features of the building and the direction and amount of driving rain are specified, and calculate the amounts of water absorbed into the building's surface when the absorption capacity of each material is known. Finally it should calculate the amounts of water running off the surface, and identify the points on the facade where this run-off stops.

The structure of this type of model can comprise of three parts.

a. A mechanism for developing a three dimensional model of a building, using a suitably adapted software package, e.g. a CAD package.

b. An analytical engine where required data describing the topography of the building or parts of the building are combined with input data about the materials and its environment. Through calculations, the nature of likely weathering effects on the modelled surfaces is determined.

c. A visualisation of the weathered facade, based upon rendering techniques applied to the original model (Figure 1).

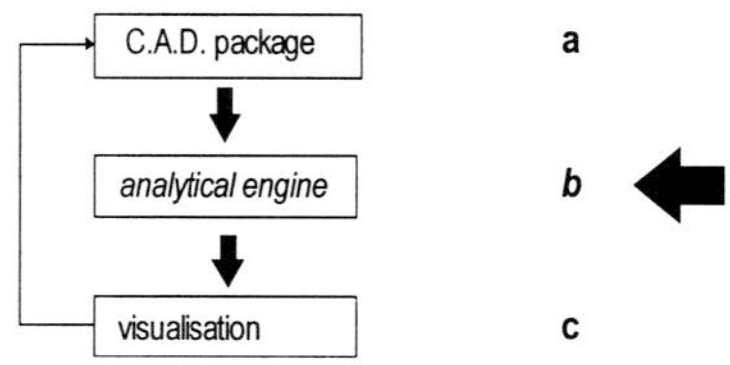

Figure 1 Parts of the model

This research is interested in the *engine* part (b) of the model described above. It addresses the most fundamental part of the problem which relates to the prediction of rain run-off.

The data needed for the development of a model for predicting weathering can be divided into three categories:

- data concerning the dimensions, geometry and orientation of the building
- data describing the materials used, especially absorption capacity
- data describing the characteristics of the environment where the building will be built, i.e. the amount of rain falling in the area, the direction and speed of wind, pollution.

Proposed Modus Operandi: Analytical Engine

The method adopted in this study for modelling water movement on a surface consists of the calculation of the amounts (volume) of rainwater running over each area of the building. The aim is initially to identify the areas and points where proportionally more water accumulates over a certain period of time.

A suitable technique is required to calculate the amounts of water on each part of the modelled surface. Finite Element Methods are often used by engineers for structural analysis. The basis of these methods is that the body or structure, in our case the building facade, is represented by an assembly of subdivisions called *finite elements*. The facade is thus divided into smaller areas, connected at certain points, called nodes or nodal points. Instead of solving the problem for the entire body of the facade, it is solved for each constituent (8, 9). If the facade is broken down into much smaller areas, the complexity of the problem is much smaller and thus easier to resolve. After carrying out calculations within each finite element, the information is joined together to form the general picture of the facade. Going from part to whole should make it possible to predict the weathering patterns on the whole facade.

Choice of finite element dimensions and shapes depends upon the size of the facade, the variety of materials used and the geometry of the building. The relationships between neighbouring finite elements may also be determined according to the materials and the geometry used on the perimeter of each finite element. The properties of the materials, e.g. absorbency, texture, influence the relationship between neighbouring finite elements. Once a building is modelled and the model divided into finite elements, a calculation procedure can be instigated (Figure 2).

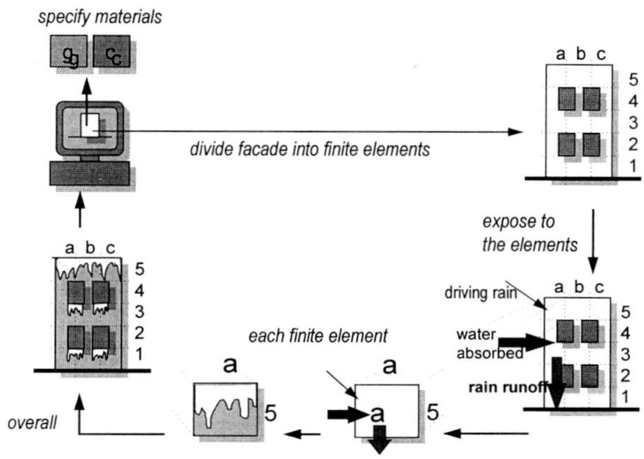

Figure 2 Analytical engine

It is proposed that this should be based upon standard weather years applied to the building. The total amount of water running over each finite element of the building can be calculated in relation to rainfall, as well as the total amount of water absorbed at each of the building's elements. From this, by reference to soiling data, weathering patterns may be theoretically predicted. This research is principally concerned with the quantification of run-off and its role in determining soiling patterns. Although the proposed model could theoretically be applied to any building material, concrete seems to provide a good case for investigation, because weathering on concrete surfaces is often negatively commented on and soiling problems, mainly caused by water, are very clearly and easily identified for modelling purposes.

Development of the Model

The proposed analytical engine has two components: the input component, where general information and data concerning the building are entered, and the analysis and calculation component where calculations of rain run-off are carried out to determine wetting patterns and water movement on the facade.

The input component includes defining the building's orientation, acquiring wind and rain data for the area and dividing the surface into finite elements according to absorption coefficients of materials and building geometry.

The analysis and calculation component includes calculating the amount of rainfall impinging on each finite element, calculating time to reach surface saturation for each finite element, calculating the amount of water absorbed into each finite element until surface saturation and finally calculating the amount of rain run-off created on each finite element. These calculations are applied to all finite elements on the building facade from the top of the facade until the ground level. The formulae used for the calculations derived from thorough analysis of existing research on driving rain, rainfall distribution, and water absorption mechanisms (10-16).

Due to the complexity of the problem, certain assumptions have been made for purposes of the initial experimental stages of the model.

- The materials of the building's surface are initially considered to have the same absorption capacity / sorptivity throughout their width. The materials are also considered to be evenly wet when the first amount of rainfall impinges on the surface.
- The time taken for materials to reach saturation refers to time taken for *surface* saturation.
- Rainfall during each day is considered *continuous*. Evaporation taking place in between rainfalls will not be taken into account in this research.
- The direction of wind is considered *constant*, i.e. there is no variation during each rainfall.
- In the first instance the surrounding environment, i.e. trees, buildings, will not be taken into consideration, although these obstacles influence the wind patterns around the building. The initial samples were considered to exist in an open area, free from any obstructions.

In the future, it would be desirable to add further parameters, e.g. layered sorptivity, surrounding environment, temperature, pollution, which can make the model more refined and closer to reality.

Validation

There were two ways by which the model could be validated. One was by comparing the results with those derived from another similar model, and the other was by empirically comparing the results of the model with real examples. Since there is no other similar existing model, empirical methods must be adopted. The model was thus applied on a number of existing buildings, and the results were compared to their present wetting and weathering patterns in order to determine the degree of accuracy of the proposed model.

One of the case studies was part of the facade of a concrete building in Blackfriars, Southwark, London. The criteria for choosing this building was its surface texture, its existing weathering patterns, its orientation and geometry.

On this building the existing staining patterns were evident and accentuated. It was thought that it could be possible to identify and predict the long-term effects of rain run-off by adding up the number of days rain run-off is produced on each finite element. This method is based on the notion that weathering patterns are a result of the accumulative action of rain run-off. Thus the finite elements which receive more days of rain run-off are likely to be less soiled than those who receive fewer days of run-off.

The facade was modelled for one whole weather year and the finite elements with more days of run-off were assumed cleaner than the ones with less run-off.

Blackfriars (Southern facade)

number of days when run-off is created on each finite element

	A	B	C	D	E	F	G	H	I	J	K	L	M	N	O	P	Q	R	S	
34	30	30	30	30	30	0	0	0	17	18	19	26	28	30	30	30	30	30	30	34
33	30	30	30	28	26	16	16	16	16	17	16	17	17	27	25	30	30	30	30	33
32	30	30	28	26	19	58	58	58	16	58	58	58	58	18	58	58	58	58	30	32
31	30	30	27	23	17	58	58	58	14	58	58	58	58	18	58	58	58	58	30	31
30	30	28	26	19	16	41	41	41	14	40	40	40	40	18	40	40	40	40	30	30
29	30	28	23	17	16	41	41	41	14	36	36	37	38	18	38	37	37	37	30	29
28	29	27	20	17	16	58	58	58	9	37	8	58	58	58	58	9	3	3	29	28
27	28	27	19	16	16	58	58	58	3	39	8	58	58	58	58	9	0	0	28	27
26	28	26	19	16	16	38	38	38	3	38	8	38	38	38	38	8	0	0	28	26
25	28	25	19	16	16	37	37	37	3	35	8	31	31	31	31	8	0	0	28	25
24	28	23	17	16	15	58	58	58	3	35	8	58	58	58	58	7	0	0	28	24
23	28	23	17	16	15	58	58	58	3	35	7	58	58	58	58	7	0	0	28	23
22	28	23	17	16	15	36	36	36	3	35	7	31	31	31	31	7	0	0	28	22
21	27	20	17	16	15	36	36	36	3	33	7	31	31	31	31	7	0	0	28	21
20	27	58	58	58	15	58	58	58	3	33	7	58	58	58	58	7	0	0	28	20
19	27	58	58	58	15	58	58	58	3	33	7	58	58	58	58	7	0	0	28	19
18	27	30	30	30	15	1	1	1	3	31	7	31	31	31	31					18
17	27	21	17	16	15	1	1	1	1	31	7	31	31	31	31					17
16	27	20	17	16	15	58	58	58	58	31	7	58	58	58	58					16
15	26	20	17	16	15	58	58	58	58	31	7	58	58	58	58					15
14	26	20	17	16	15	13	13	12	9	31	7	31	31	31	31					14
13	26	19	17	16	15	8	8	8	4	31	7	30	30	30	30					13
12	26	19	17	16	15	58	58	58	58	31	7	58	58	58	58					12
11	26	19	17	16	15	58	58	58	58	31	7	58	58	58	58					11
10	26	19	16	16	15	8	8	8	6	31	7	30	30	30	30					10
9	26	19	16	16	15	3	3	3	2	30	7	30	30	30	30					9
8	26	58	58	58	15	58	58	58	1	30	7	58	58	58	58					8
7	26	58	58	58	15	58	58	58	1	30	7	58	58	58	58					7
6	26	19	16	16	15	3	3	2	1	30	7	30	30	30	30					6
5	25	19	16	16	15	3	2	1	1	30	7	30	30	30	30					5
4	25	19	16	16	15	3	2	1	1	30	7	58	58	58	58					4
3	23	19	16	16	15	3	2	1	1	30	7	58	58	58	58					3
2	23	19	16	16	15	1	1	1	1	30	7	30	30	30	30					2
1	23	19	16	16	15	1	1	1	1	30	7	30	30	30	30					1

Figure 3 Visualisation of calculation results

Figure 4 Existing weathering patterns

Figure 3 was the result of these calculations and when compared to Figure 4 there seems to be reasonable correlation. The points where there is least correlation is likely to be due to the method of estimating rainfall distribution and the large size of finite elements.

CONCLUSIONS

The validity of the principles followed for the development of a predictive model for weathering seem to be reasonable. The case studies used for validation showed a general correlation between existing wetting / weathering patterns and predicted wetting patterns.

Further development and refinement are naturally essential for the production of a package which could help designers optimise their work with regard to prevention of unsightly weathering.

REFERENCES

1. MOSTAFAVI, M., LEATHERBARROW, D. On Weathering; the Life of Buildings in Time. (1993). M.I.T. Press. Cambridge. Massachusetts.

2. BAKER, M.C. Rain deposits, Water Migration and Dirt Marking on Buildings. RILEM / ASTM / CIB Symposium on Evaluation of the Performance of External Vertical Surfaces of Buildings. Otaniemi. (1977). Vol. I. pp. 57-67.

3. ATKINSON, B. Weathering and Performance. Weathering and Performance of Building Materials. ed. Simpson and Horrobin. MTP. (1970). Aylesbury. pp. 1-40.

4. VERHOEF, L.G.W. (ed) Soiling and Cleaning of Building Facades. RILEM Report. Chapman and Hall. (1988). London. p. ix

5. C.I.B. Quantities, Symbols and Units for the Description of Heat and Moisture Transfer in Buildings; Conversion Factors. Conseil International du Batiment pour la Recherche. Rotterdam. (1977). Report No.37.

6. BEIJER, O. Weathering on External Walls of Concrete. Swedish Cement and Concrete Research Institute. (1980). Stockholm. pp. 21-24.

7. HAWES, F. Appearance Matters 6; The Weathering of Concrete Buildings. Cement and Concrete Association. Wexham Springs. Slough. (1986).

8. PICKARD, A. C. The Application of 3-Dimensional Finite Element Methods to Fracture Mechanics and Fatigue Life Prediction. (1986). E.M.A.S. West Midlands. U.K. p. 50.

9. DESAI, C., ABEL, J. Introduction to the Finite Element Method; A Numerical Method for Engineering Analysis. Van Nostrand Reinhol Company. .(1972). New York. p. 3.

10. LACY, R. E. Climate and Building in Britain. B.R.E. report. H.M.S.O. (1977) London.

11. ZHU, D., FAZIO, P., MALLIDI, S. R. A Qualitative Study for the Measurement of Driving Rain Exposure for the Montreal Region. Building and Environment. (1995). Vol. 30. No. 1

12. CHOI, E.C.C. Velocity and Impact Direction of Wind Driven rain on Building Surfaces. ICBEST 97 Conference Proceedings. Bath. U.K. (1997) pp. 465-472.

13. HALL, C., KALIMERIS, A. N. Rain Absorption and Runoff on Porous Building Surfaces. Canadian Journal of Civil Engineering (1984). Vol. 11. pp. 108-111.

14.HALL, C., YAU, M. H. R. Water Movement in Porous Building Materials; IX The Water Absorption and Sorptivity of Concrete. Building and Environment (1987). Vol. 22. No. 1. pp. 77-82.

15.ELSHIMI, M., WHITE, R., FAZIO, P. Influence of Facade Geometry on Weathering. Canadian Journal of Civil Engineering (1980). Vol. 7. pp. 597-613.

16.GUMMERSON, R. J., HALL, C., HOFF, W. D. Water Movement in Porous Building Materials; II. Hydraulic Suction and Sorptivity of Brick and Other Masonry Materials. Building and Environment (1980).Vol. 15. pp. 101-108.

RAPID HARDENING DECORATIVE COMPOSITIONAL PORTLAND CEMENT WITH MECHANICAL-CHEMICAL ACTIVATION

M Sanitsky

G Shevchuk

M Loza

R Soltysik

Lviv Polytechnic State University

Ukraine

ABSTRACT. This paper presents results of studies on decorative binding materials made using white rapid-hardening composite Portland cement containing alkaline and mineral additives. The influence of mechanical-chemical activation of decorative binders by means of vibration mills on their structure formation and physical-mechanical properties of mortars and concretes is established. The decorative water-resistance gypsum-cement-pozzolanic binders are developed using white rapid-hardening composite Portland cement. Investigation of complex chemical admixtures influence on the setting time of gypsum is given.

Keywords: White ordinary portland cement (WOPC), Additives, Activation, White rapid hardening compositional portland cement (WRHCPC), Multicomponent gypsum binders, Decorative gypsum-cement-pozzolanic binders.

Professor Dr Sc Myroslav Sanitsky is the head of the division "The Technology of Binding Materials (Cements)", Lviv Polytechnic State University, Lviv, Ukraine. He specialises in the use of binders and the durability of concretes. Professor Sanitsky has published widely and is an Academician of the Academy of Civil Engineering of Ukraine.

Dr G Shevchuk is an Assistant, Chair of The Chemical Technology of Silicates, Lviv Polytechnic State University, Lviv, Ukraine. His main research interests include the permeation properties of cements, and the investigation of cement hydration processes.

Dr M Loza is a Research, Chair of The Chemical Technology of Silicates, Lviv Polytechnic State University, Lviv, Ukraine. His main research interests include the investigation of decorative composite portland cements with mechanical-chemical activation.

R Soltysik is a post-graduate student, Chair of The Chemical Technology of Silicates, Lviv Polytechnic State University, Lviv, Ukraine.

INTRODUCTION

The effective finishing materials, based on decorative cements, play first fiddle in the modern construction. These cements are white Portland cements as a rule. Its production technology causes heightened heat and energy losses because of features of chemical composition of raw mixture. This situation causes an increase in finishing materials cost [1, 2]. The shortening of materials and energy losses during finishing of construction is possible at the expense of decorative composite cements containing mineral additives. The composite Portland-cements and cements contain comformably 6-35 mass.% and 36-80 mass.% of active mineral additives and fillers according to standard ENV-197 [3]. Thus, traditional composite cements are characterised by kinetics moderation of strength acquire, therefore it is necessary to use different activation methods.

Mechanical and alkaline activation seems to be the best way to improve the Portland cement binders efficiently. The mixing and milling of white Portland cement with alkali-containing additives, mineral additives, and also pigments, with suitable quality characteristics in the vibratory mill is an optimum manner for decorative cements improvement.

This article is devoted to investigation of mechanical-chemical activation influence of decorative cements with mineral additives of light colours on their parameters. The parameters are connected with fineness of grinding and granular characteristics. Also the paper is devoted to studying hydration and hardening of the Portland cement systems. It permits to develop rapid hardening decorative composite cement and finishing materials with determined constructional and technical properties.

EXPERIMENTAL DETAILS

Materials

White ordinary Portland cements (WOPC) and white rapid-hardening composite Portland cement (WRHCPC) were used. Alkali containing waste of glass batch that have calcium carbonate and sodium sulphate and carbonate was used as a hardening accelerator and a filler.

Equipment

Vibration mills with a volume of 25-1000 dm^3 were used for grinding different binding materials. Mills with container's volume of 100-1000 dm^3 may be used in serial production. Depending on initial parameters of material their productivity is 100-3000 kg/h. The container is set on the bed (plate) with the help of pneumatic balloons, which ensure practically full vibroisolation. Vibrative mills of this type with the volume till 25 dm^3 we used in laboratory conditions for researching the influence of grade on qualitative characteristics of materials and their compositions. By small power inputs (capacity - 30 kW) in the regime of grinding the productivity of vibrative mill MV-1000 reaches 3-5 t/h and in the regime of finishing grinding and homogeneous mixing - 6-9 t/h.

RESULTS AND DISCUSSION

The possibility to use of heightened quantity of the mineral additives as blust furnace slag, zeolite rock, silica clay, quartz sand, limestone and chalk for obtaining decorative binder based on white Portland-cement is investigated. It is determined, that activity of binder is increasing during mixing of finely dispersed additives with white Portland-cement and causes the losses of decorative mortar. Besides, the limestone is the most effective mineral additive, because its use in quantity of 20 mass.% leads to insignificant decreasing of strength of cement paste.

The main property of white Portland-cement is degree of whiteness, that determines the quality of decorative materials. The addition of mineral additives to white Portland-cement leads, as usual, to worse binder whiteness. However, the replacing of part of white Portland-cement on chalk leads to increasing of binder whiteness. Thus, the use of carbonate raw materials as an active mineral additive causes a positive influence on both strength and whiteness of white Portland-cement. It is determined, that quality of white Portland-cement depends on finely dividing of components, on relation of finely dispersion and uniformity of distribution on powder and on whiteness of used additives. Therefore, white composite cement request more finely dividing than Portland-cement with the same strength class.

The investigation of effectivity of mechanical activation of compositional cements showed that specific energy losses for changing of dispersion in vibratory mill are 3-5 times lower in comparison on ball mill. The admixing of Portland-cement in vibratory mill, unlike in ball mill, to the specific surface 500 m^2/kg permits to increase the contain of finely dispersed fraction (less than 3 μm) to 2-3 times and maximum decrease the contain of particles more than 60 μm. It is important factor in the effective using of possibilities of white Portland-cement clinker.

The amount of C_3A increases in composition of finely dispersed fraction during admixing of white Portland-cement, as a result of features of its mineral composition (C_3S =45,37 mass. %; C_2 S = 35,43 mass. %; C_3A = 15,47 mass. %). Therefore the cement is characterized by acceleration setting because additive of gypsum in quantity 5-6 mass.% can bind only 3,0 mass.% C_3A. X-ray analysis showed that prehydration phases of white Portland-cement are calcium hydroxide, ettringite and hydroaluminates C_4AH_{13-19}. The transformation of hexagonal hydroaluminative A_m-phases into cubic C_3AH_6 is the cause of decreasing of white Portland-cement strength. It accompanied by increasing of system porosity to the 20-30% as a result of different compactness of hydrates and makes negative influence on properties of white Portland-cement. The necessity of stabilisation of hexagonal A_m-phases is growing up as a result. It is possible to achieve this by means of carbonate calcium, which is interacting with hydratation products of C_3A and create the hydrocarboaluminate calcium $C_3A \cdot CaCO_3 \cdot 12H_2O$. The crystal-chemical analyse of structure of hydratical phases was made and modificative role of carbonate additives during sructurization of composite white Portland-cement was shown at the examples of model binders C_3S + C_3A. The structure-active hexagonal phases are stabilising at the expense of creating of hydrocarboaluminates calcium. It leads to the increasing of strength of cement stone.

The effectivity of mechanical-chemical activation into vibratory mill of white Portland-cement with alkaline and mineral additives vibratory mill is shown as a result of investigation of influence of mechanical-chemical activation on properties of white Portland-cement. The complex alkaline- and carbonate-containing additive CACA (waste of glass batch) show properties of intensificators of process of mechanical activation and hardening accelerator. It permits to obtain rapid-hardening white Portland-cement.

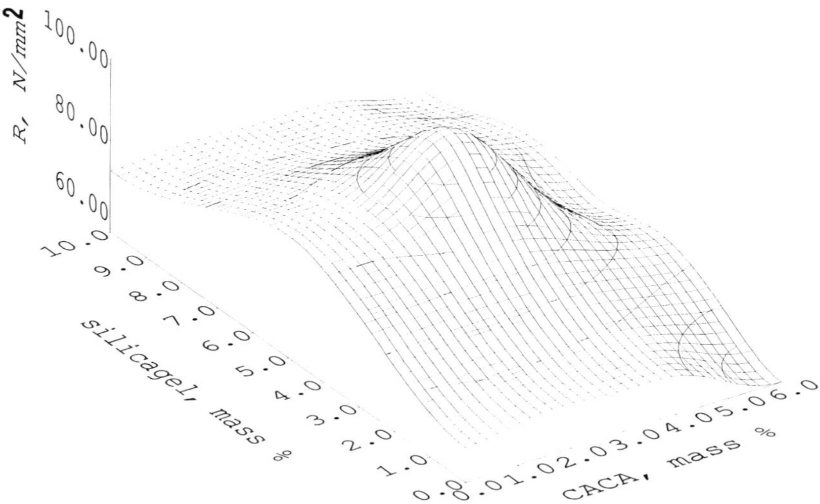

Figure 1 Isolines of egual compressive strength WRHCPC depending on quantity of additives at 28 days

White Portland cement is characterised by a low value of early strength of cement stone. An increase in specific surface of white Portland cement to 560 m^2/kg at the expense vibro-activation permits to increase early strength of cement during addition of mineral additives to the contain 40,0 mass.%. The decreasing of strength is observed during 7 days of hardening in comparison of usual white Portland cement. It means that cement of 28 days age hardening is characterised by a low value of strength. The use of finely dispersed waste of glass production and silicate gel permits to obtain the specific surface 536 m^2/kg and increase early strength of cement, when the energy losses are two times lower. This cement complies with requirements of rapid hardening white composite Portland-cement M400 according to GOST 310.4-81 (Table 1) and CEM II/B-M 32,5R according to ENV-197. The additions to the contain of white composite Portland cement the superplasticizer S-3 permits to obtain binder with low W/C, which is characterised by high early strength in comparison with usual white Portland cement.

The necessity of expansion of base for producing of finishing materials make possible to use for architectural-finishing works not only white but usual Portland-cement. The producing of decorative binders of middle and light colours is possible on base of gray Portland cements.

It is determined, that using of mechanical activation of usual Portland cements and mineral additives of ligth colour (limestone, chalk, zeolite) into vibratory mill provide the improved decorative properties and saving of 40% of pigments during obtaining of decorative composite cements.

It is possible to use both rapid hardening decorative composite Portland cements and multi-components modificated binders, based on hemihydrate gypsum during finishing works. The usual constructional gypsum is characterised by low water-resistance, short period of setting, therefore it is not possible to use it for finishing works. The improving of technical properties of gypsum binders achieve at the expense of their modification by mineral and chemical additives during mechanical activation into vibrative mill.

Table 1 Physical-mechanical properties of cements according to GOST 310.4-81

CEMENT	W/C	SA, m^2/kg	WH,%	SETTING, h-min: Initial / Final	STRENGTH, N/mm^2 compressive / bending: 20 °C, 2 d	20 °C, 3 d	20 °C, 28 d	after HC, 1 d
WOPC	0,29	310	70	2 – 00 / 2 – 50	8,2 / 1,2	9,8 / 1,6	30,1 / 4,7	21,1 / 4,0
WRHCPC	0,26	424	77	1 – 00 / 1 – 30	20,0 / 3,9	26,9 / 4,6	45,3 / 6,6	27,7 / 5,0

* SA=surface area WH=whiteness HC= heat curing

The decorative gypsum-cement-pozzolanic binders are developed with using of white composite cement. They are characterised by heightened water-resistance, high early strength and necessary periods of setting. The additives of pozzolanic action and additive for adjusting of hardening period were used for production of modificated gypsum-cement-pozzolanic binders. The increasing of acquire strength intensification is observed during adding the mineral additive zeolite to contain the decorative binder. The water resistance increasing to 22% as a result of interaction between mineral additive and calcium hydroxide, and formation of calcium hydrosilicate, which are promote to increasing of water-resistance of units.

Mechanical activation of multi-components modificated gypsum binders into vibration mill increases their strenght, but accelerate the period of setting. The Na_3PO_4 in quantity 0,1-0,2% was used for adjusting of setting periods of decorative modificated gypsum binders.

It permits to decrease the beginning of setting to 1-2 hours. The influence of Na_3PO_4 on degree of gypsum hydration in contein of modificated decorative gypsum binder was determined by method of quantity X-ray analyses. The line of hemihydrate gypsum with d/n=0,598 nm was used an analitytical. It is determined, that 4 min after mixing multicomponents decorative gypsum binder with water, the degree of gypsum hydration is 50%, after 15 min - 98%. The phase composition of stone do not change during of addition 0,1 mass% Na_3PO_4, but deceleration of hydration process of hemihydrate gypsum is observed. So, the degree of hydration of hemihydrate gypsum after 8; 15 and 30 min is 5; 9 and 50%. The full hydration occure after 70 min. The additive Na_3PO_4 increase the process of solution of hemihydrate gypsum, according to the given results. The film is creating on the surface of gypsum grains as a result of interaction between gypsum with Na_3PO_4. This film decelerates the hydration process. The duration of retardation hydration process depends on thickness of film. The thickness of film depends on Na_3PO_4 concentration and specific surface of binder. The film thickness and acceleration of setting period of multicomponents modificated gypsum binder are decreasing because the specific surface increasing during mechanical activation.

PRACTICAL APPLICATION OF RESULTS

The expediency of using of rapid hardening white composite Portland cement during production of decorative mortars for finishing works is shown. During investigation of decorative cement-lime mortars was determined, that using of decorative mortars, based on rapid hardening white composite Portland cement, increase the strength and frost resistance of finishing mortar, permits to decrease the loss of binder to 35% on 1 m^3 of mortar. The decorative concretes, based on the white rapid hardening composite Portland cement are characterised by increased strength, frost resistance and durability.

CONCLUSIONS

1. The influence of complex alkaline and carbonate containing additives and silicate gel on physical-mechanical properties of white composite Portland cement is investigated. It is determined, that use of mechanical activations of composite Portland cement by means of vibratory mills permits complex alkaline activation to accelerate the hydration processes at early age, and for binding material with increased strength (up to 1.5 times) to be obtained.

2. The processes of structure formation of white rapid hardening composite Portland cement were studied by means of complex methods of physical-chemical analyses. It was shown that alkaline components leads to an increase in alite phase hydration degree and that calcium carbonate modifies the hydration products of C_3A to create hydrocarboaluminate $C_3A \cdot CaCO_3 \cdot 12H_2O$ which is structure-active component of Portland cement paste.

3. Decorative multicomponent modificated gypsum binders with increased water resistance and regulated time of setting are developed.

REFERENCES

1. SANITSKY, M A, SHEVCHUK G J, LOZA M Z AND SMITSNIUK R V. Influence of alkali metal salts on formation of structure of portland cement stone. Proceedings of the First International conference on alkaline cements and concretes. Kyiv, 1994. pp 303-311

2. SANITSKY, M A, MELNYK, V M AND CHABA, P M. High early strength portland cement binders for using at low ambient and elevated temperatures. Proceedings of the International Conference held at the University of Dundee, June 1996, pp 649-654.

3. GOLOVANOVA, S.P, ZUBEHIN, A.P AND KOGENTSEV, U.T. Decorative silicate materials in architectoral design . Cement and its using., No 2, 1997, pp 31-32.

4. SCHULZE, W, TISCHER, W AND ETTEL, W. The mortars and concretes on noncements binders. Translated from Germany. Moscow, Stroyizdat, 1990, 240 p.

IMPROVING THE QUALITY OF LIFE WITH URBAN CONCRETE ARTWORK

P Neal

J Clifford

Cape Technikon

South Africa

ABSTRACT. Artistic talent abounds in people of all walks of life. Many materials have been used throughout history to express these talents but few have the lasting qualities of concrete. Concrete however, has a poor public image, being considered the grey, cheap industrial construction material. To date it has not become a medium for the average citizen to use in enhancing private property or residential streets. Research at the School of Civil Engineering to improve the quality of life in urban residential streets uses concrete as a functional and artistic medium for this purpose. This research led to a revolutionary concept known as "Street Bodies Corporate", whereby the ownership and management of residential streets is by residents rather than municipalities. This paper considers the useful contribution of artistic concrete in job creation and to enhance the quality of life in such urban residential communities. Concrete is used for segmental block paving, streetlights, boundary walls, tree surrounds, seating arrangements and rubbish containers.

Keywords: Civil engineering, Community, Concrete, Job creation, Quality of life, Residential, Urban.

Patrick Neal is a technologist with the Cape Technikon, South Africa, carrying particular responsibility for the practical elements in civil engineering. He is also actively involved in stimulating secondary school students to consider civil engineering through industry-supported schemes. He is involved in an extensive research programme to improve the quality of life in urban residential communities. He is married to Christiana and the couple have two children. As an avid golfer Patrick hopes to realise a lifetime ambition to play at St Andrews.

Dr John Clifford leads the Street Bodies Corporate research project at the Cape Technikon. He also facilitates students towards self-actualisation in tertiary learning, having pioneered the School's new research culture. He is an innovator and his doctorate from Pretoria University, South Africa, was awarded for research leading to the use of concrete segmental paving for the heaviest loading applications. He is married to Fiona and the couple have three married daughters and four grandchildren.

INTRODUCTION

The South African government is hard-pressed in providing enough housing for all its people. Whilst much has been accomplished in providing housing and essential services, the political climate is affected by the frustration of the significant number of people who have not been provided with housing. South African government ministers, as in many other developing nations, have limited resources and might seek desperate solutions. With the South African economy performing slower than expected (predicted not to exceed 3 % growth in 1999), no major breakthrough can be expected for those people who are desperate for basic housing.

Clearly, in South Africa and in many other parts of the world, the need for the provision of basic housing must be addressed. Practical and constructive solutions for cheaper construction methods to serve the ever increasing population growth in the developing world must be found. Unfortunately, politicians have sometimes created unrealistic expectations in the poorest members of the electorate, trading votes for the hope of housing and services. Lower-income groups of people cannot afford the luxury of living in clay brick houses built by developers and cheaper alternatives must be found. Some solutions have been found. Various alternatives are possible including adobe structures as found in east Africa and elsewhere, and the meagre and primitive mud and straw huts in central Africa [1]. The former clearly has great potential, especially in Southern Africa, and research has established its viability [2]. The latter material though is obviously a more temporary solution because of its limited ability to withstand the extremes of the African climate.

But concrete, comprising elements which can be obtained and processed by self-help labour-intensive activity, suggests a stable and viable option to the housing dilemma. This versatile product can readily be used to provide long-term solutions to the problems of how to provide a rapid solution to the housing problem in Southern Africa. And not only can self-help concrete provide the basic shell structure of houses and other needed buildings, but it can also be shaped and finished in a variety of ways, creating original form and expression.

THE SOUTH AFRICAN EXPERIENCE

In South Africa concrete has traditionally conveyed a poor public image. For many it is conceived as grey, mundane, lifeless, and unattractive. Yet the product's potential is so different to this confined image. Increasing numbers of important buildings are being made of concrete with a variety of textures, finishes and pigmented colours. There are many examples of such enhancement, especially within the Cape Town CDB, some structures displaying unusual artistic flair and ingenuity. This paper suggests that the flair and artistic creativity seen in prestigious buildings can be used in self-help solutions to the housing problem.

Research at the Cape Technikon's School of Civil Engineering has led to a revolutionary concept known as "Street Bodies Corporate [3]. The concept encourages existing and proposed communities to design, construct, and maintain the services located in residential streets. In addition, by the planting of trees and other flora potential resources for job creation are provided whilst enhancing the quality of life of the residents. Other aspects of Street Bodies Corporate include the option for residents to clean their own streets, supplement their own electricity supply, harvest rainwater, and convert wastes into methane and fertiliser. By these means not only are jobs created close to home but the municipal rates can be reduced. Naturally, self-help concrete artefacts make a useful contribution to the goals of Street Bodies Corporate.

INNOVATION FROM THE PAST

Thomas A. Edison, the so-called "Wizard of Menlo Park", did not confine his studies to the light bulb for which he is principally remembered. He believed that concrete provides a low-cost method of house construction and proved his point by the mass production of cheap concrete homes. He wrote:

> "I have conducted a large number of experiments. These experiments have proved that it is possible to cast a house complete in six hours by pouring a very rich mixture of sand, gravel, and cement into moulds ... the hardening of the concrete will require four hours. Well made concrete, employing a hard grade of Portland cement is the most lasting material. The iron moulds for the full size house are at the present time [circa 1908] about 60 % complete and it may be possible that before the fall they will be finished and the first house cast. These houses will be waterproof and damp proof. After the forms are removed the houses can be painted with a paint made of cement tinted with red oxide which hardens and never deteriorates ... "I think the age of concrete has started and I believe I can prove that the most beautiful houses that our architects can conceive can be cast in one operation in iron forms for a cost which will be surprisingly low. Then even the poorest man among us will be enabled to own a house all his own and a house that will last for centuries with minimum cost for insurance or repairs." [4]

It is with the above kind of innovation and bold thinking that the housing dilemma in many underdeveloped countries of the world can be resolved. Edison also cast all the furniture in some of his concrete houses in concrete. The furniture even included a beautiful concrete phonograph cabinet which can be seen in Figure 1. Edison's innovate concept of self-help concrete need not be limited to the provision of basic housing and furniture but can apply to street amenities which can be cast either in-situ or in moulds, special attention being given to aesthetic finish and colouring.

CONCRETE HOMES FOR THE MODERN DAY

A more recent innovative use of concrete for housing, known as Lightweight Composite Concrete (LWC), was developed by John August [5]. Whist the prototype was designed for the needs of Hawaii it can be adapted to suit any climate. Embodying many architecturally unique design features, the structural component of LWC is fascinating. Steel reinforced concrete shear columns support a monolithically cast LWC curvilinear roof and ceiling system giving the dwelling a cool ambiance due to high-insulating qualities.

As can be seen in Figure 2 rainwater naturally flows to the corners of the structure where it can be directed to a storage tank. The eaves overhang by 2 m and the upper balcony is fully covered. The total building, which is 6 m high, is claimed to be fire, rot, termite, and hurricane proof and can apparently withstand a moderate earthquake without structural failure.

According to the author (J August) , complete plans will soon be available for developing the concept for use in low to moderate cost housing. As with Edison August has manufactured modular concrete furniture. Figure 3 shows a terraced dining patio adjoining the prestigious Kula Lodge on Maui, Hawaii, which required special seating criteria; resistance to adverse weather, low maintenance, and permanency. The interlocking modular system comprises five parts, enabling each bench to be easily assembled on site. The moulds used were a hybrid composite of polyurethane rubber on the inside bonded to an outside shell of epoxy-fibreglass-carbon fibre.

Figure 1 Edison's concrete phonograph cabinets (Courtesy of Cement & Concrete Institute Library, Johannesburg)

Figure 2 Artists' impression of a concrete home in Hawaii

JOBS CAST IN CONCRETE

Street Bodies Corporate require residential streets to become *cul de sacs* wherever possible and by stopping through traffic the speed of vehicles in residential streets is reduced considerably. Concrete segmental paving provides an aesthetic practical alternative to "black top" surfacing but fails to provide the optimal "green" trafficable surfacing. A new trafficable segmental self-help concrete paving system called Enviropaving was developed to serve these needs. The system is similar to other green blocks but maintains an uncompacted agricultural soil regime in which grass and other ground cover thrives. Traffic loads are transmitted from surface concrete ribs by means of lateral support strip beams to a structural support layer comprising compacted earth. A prototype of the system at the Du Noon informal settlement near Cape Town is performing satisfactorily under limited traffic whilst controlling rainfall runoff. The ground cover not only enhances the visual aesthetics but provides food for chickens and goats, typical paver designs shown in Figure 4.

MODERN DAY PRACTICALITIES

As cities expand cemeteries and graveyards in which the remains of those long forgotten lie seem sacrosanct limiting further development. Clifford has proposed the exhumation of the remains for cremation of mass burial elsewhere prior to making these once important areas again useful to the living [6]. Where religious practice does not allow such treatment he proposes the areas be planted with indigenous trees amongst which wildlife can be introduced as an urban park. However, disposal of the ashes of the dead within Street Bodies Corporate suggests the optimum solution. The remains are placed in artistic self-help concrete tree surround benches as shown in figure 5. These benches represent a practical alternative to the traditional crematorium "hall of remembrance". The benches are formed from precast elements that, when placed together, encircle an indigenous shade tree.

Figure 3 Modular interlocking concrete seating

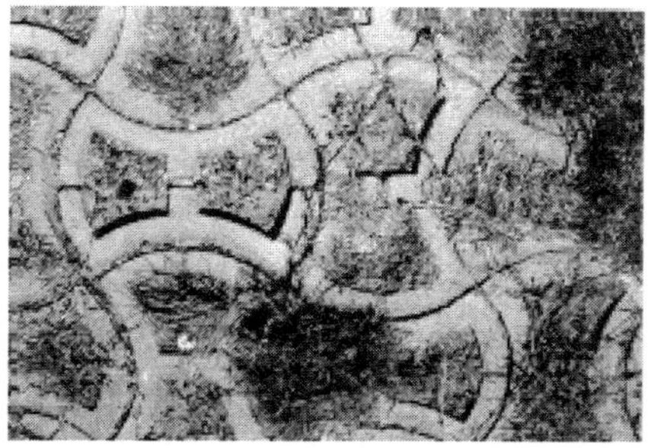

Figure 4 Typical interlocking concrete paver designs

Numerous other self-help functional and artistic concrete artefacts can be envisioned to enhance Street Bodies Corporate. These include concrete bollards with built-in light fittings, curbs, segmental plantable (green) garden walling units, end-of-street postboxes and rubbish containers. Self-help concrete can be provided with attractive exposed-aggregate surfaces to blend in with the green environment. Many mass-produced lamp poles are functional but unattractive. Self-help concrete provides the opportunity to recapture a less-stressful by-gone era. Each could be individually designed to include crossbars upon which hanging baskets full of flowers can be placed.

This type of lamp post would be especially suitable for use in communal city gardens and squares. Their manufacture could provide jobs in an entrepreneurial back-yard industry.

CONCRETE TO ENHANCE GARDENS

Gardens add to the quality of life for they provide a peaceful location in which to relax. Street Bodies Corporate envision each residential street as a garden serving as a safe play area for children, a place of relaxation for the sick and elderly, yet a place to park vehicles overnight. Self-help concrete artistic artefacts create job opportunities whist adding individuality to both Street Bodies Corporate and Corporate and communal city garden areas. The list of concrete artifacts to grace gardens includes statues, fountains and other water features, paved adventure paths for small children, sundials, moveable planters, and fencing and screening.

In terms of innovation and personalisation a recent competition by the School's concrete technology students might be mentioned. As part of a practical hands-on and team-building exercise a competition with prizes provided by the industry was devised. Groups of students were required to design, build and race concrete wheelbarrows.

Although the prime concept was to incorporate structural design principles with aesthetic and functional capability, several of the wheelbarrows were so original, unusual, and attractive that they were immortalised as planters and birdbaths in a garden (Figure 6).

Figure 5 Tree surround benches for crematorium ash

Figure 6 Concrete wheelbarrows converted to planters and birdbaths

By similar means urban residential areas can be personalised with self-help concrete artefacts. Traditional, city parks and other public areas have been provided with wooden seating and steel cage waste bins. These are items that could easily be manufactured in self-help concrete as an entrepreneurial job-creation exercise. Street Bodies Corporate and local parks could incorporate adventure locales to amuse and teach beneficial principles to children. They might be conceived as being made not only aesthetically attractive but incorporating places in which children can dream and have fun. Imagine an attractive self-help concrete elephant also designed to function as a waste paper bin. Children could be encouraged to dispose of their wastes by placing waste paper in the mouth or ear of the elephant. Other self-help concrete animals can be conceived, encouraging children to climb on their backs. A group of self-help concrete horses could be located behind a stand of shrubbery apparently ready to charge. Giraffes could be placed so that they peek above trees. The back of a bush pig would make an unusual place for children to ride on and they could hold onto its large front teeth. The animals of the African jungle could be reintroduced where they used to roam, at least conceptually by means of self-help concrete artwork. How much would this add to the fun of a childhood even within the densely populated urban residential streets, transposed through the concept of Street Bodies Corporate.

CONCLUSIONS AND RECOMMENDATIONS

This paper has shown briefly how self-help concrete can be used to enhance the quality of life as an element in the transposition of urban residential streets through Street Bodies Corporate. Each person has some natural talent in art and the concept allows street communities to incorporate these latent abilities in the improvement of their own quality of life. Individuals need to be empowered to break the traditional stranglehold imposed by remote often disinterested and underfinanced municipalities. When this happens urban residential streets can provide many jobs close to home, be used to grow renewable green resources, and help control urban pollution. Of greatest significance, Street Bodies Corporate with their emphasis on self-help can utilise the most universal and available self-help construction material, concrete, in overcoming the housing problems extant in many countries of the world.

REFERENCES

1. IFEOMA ONYEFULU: Daki (mud houses in Northern Nigeria and the Niger Republic. The Architectural Review (ISSN:0003-861). May 1988, pp 92-93

2. LILLEY, D M AND ROBINSON, J. Ultimate strength of rammed earth walls with openings. Academics from the University of Newcastle upon Tyre. Paper accepted for publication in the proceedings of the Institution of Civil Engineers. December 1994

3. CLIFFORD, J, LÜCH, A, SHÖNFELD, C (interns). The Urban Village. The Darwin Summit. National Engineering Conference, Darwin, Australia. April 1996..

4. THOMAS EDISON'S CONCRETE HOUSES. Concrete Construction, June 1965 Article in Concrete International Journal. August 1983, p 204

5. AUGUST, J, 1993. Addressing the need for permanent shelter in Hawaii. The Concrete House. Website on Internet. Http://www.geckostone.com/conhouse

6. CLIFFORD, J. Civilised disposal of the dead. Paper accepted for publication in Journal of South African Institution of Civil Engineers. October 1996

MAXIMISING OPPORTUNITIES FOR FABRIC THERMAL STORAGE

J Glass

C Kendrick

B Baiche

Oxford Brookes University

UK

ABSTRACT. The use of concrete's capacity to store and release heat, and thereby temper the internal environment of a building by moderating peak temperatures, can be optimised through good design and construction. As a result of ongoing research into fabric energy storage (FES), the authors argue that many opportunities exist to maximise the FES potential of concrete buildings. The paper covers both theory and practice of FES, and makes recommendations on the most effective ways in which commercial office buildings can be designed to reduce the need for air conditioning. The first part outlines the principles of fabric energy storage. Part two describes how these principles can be put into practice. Case studies of in-situ and precast concrete FES systems are included in the final section.

Keywords: Commercial buildings, Direct finish, Exposed concrete, Fabric energy storage, Precast hollowcore slabs, In-situ construction, Reinforced concrete floors, Thermal mass.

Dr Jacqueline Glass is a Senior Lecturer in Architecture at Oxford Brookes University. She specialises in tilt-up concrete, cost modelling and the architectural and cultural aspects of concrete construction. Dr Glass is the Architectural Consultant to the British Cement Association and the Reinforced Concrete Council.

Christopher Kendrick is a Researcher in the Postgraduate Research School at Oxford Brookes University. He is an environmental engineer specialising in building physics and has carried out extensive research on environmental floor systems at BSRIA and is currently completing PhD research on thermally permeable ceilings.

Dr Bousmaha Baiche is a Researcher in the Postgraduate Research School of the School of Architecture at Oxford Brookes University and provided technical assistance with this work. His research interests are earth construction, thermal mass, and hybrid concrete structures.

All are members of the Oxford Centre for Sustainable Development Technology Group.

INTRODUCTION

Many commercial buildings now have a predominant cooling requirement throughout much of the year as a result of increasing use of electronic equipment such as computers, printers and photocopiers. Air conditioning and comfort cooling is becoming significantly more common in UK offices to maintain thermal comfort. In air conditioned offices, the electrical power used for refrigeration, pumps and fans is second only to that used for lighting, and is the fastest growing consumer of electricity in the commercial/services sector. With growing concern over CO_2 emissions and global climate change, and the possible restriction on the use of air conditioning in forthcoming revisions to Part L of the Building Regulations, it seems likely that designers will look for low energy methods of achieving thermal comfort. Fabric Energy Storage (FES) uses the thermal capacity of the building fabric to attenuate and delay peak internal temperatures during the occupied period. Heat is absorbed and stored in the building structure, which is subsequently cooled by a period of night ventilation, or purging. When used effectively, FES can reduce peak internal temperature by 2-3°C, often maintaining temperatures below ambient in spite of internal heat gains [1]. Peak temperature is also shifted to later in the day, often occurring after the occupants have left. FES thus has the potential to reduce or obviate the need for air conditioning in many buildings, when used with effective solar control and ventilation. Following an introduction to the principles of fabric thermal storage, the paper discusses relevant design aspects of concrete construction and concludes with two case studies which illustrate both in-situ and precast alternatives, and passive and active design approaches.

PRINCIPLES OF FABRIC ENERGY STORAGE

Fabric energy storage uses the thermal mass of the building structure, or those elements with a significant heat storage capability, to moderate internal temperatures. Heat is transferred by convection and thermal radiation from the occupied spaces to the elements of thermal mass in the structure, where it is conducted into the element and stored until the night cooling period. A period of night ventilation reverses the process, cooling the structure for the next day. FES is ideal for commercial buildings, which have a regular cycle of daily heat gains. Any part of the structure can be used, including facades, frames, and floor slabs. The latter are particularly suitable because they form the largest volume and surface area of thermal mass and are evenly distributed throughout a building. Concrete has comparable thermophysical properties to brick or stone, but it has been found that the main constraint to effective FES is surface resistance to heat transfer rather than material properties. Three heat transfer mechanisms operate to enable fabric energy storage to occur.

Convective Heat Transfer

The transfer of heat by fluid motion is described by Newton's law of cooling:

$$q_c = h_c A(T_s - T_\infty) \tag{1}$$

where q_c = heat flux (W), T_s = surface temperature, T_∞ = bulk fluid temperature
h_c = convective heat transfer coefficient (W/m^{2}K), A = surface area (m^2)

In FES, useful convective heat transfer occurs between the air in the space and the exposed elements of structure. The temperatures can be measured easily, but the convective heat transfer coefficient is dependent on the turbulence of the flow, conditions in the thermal boundary layer (particularly influenced by the surface roughness), and properties of the fluid considered. Identification of a suitable value of h_c is thus extremely difficult, and use must usually be made of empirical correlations suitable for the conditions encountered.

Radiative Heat Transfer

The exposed underside of a cool floor slab in a room will receive energy via net radiative exchange with warmer surfaces in the surrounding room (including human occupants) and vice versa overnight. The magnitude of the exchange depends upon the temperatures of the surfaces, their areas, emissivities, and relative geometry. For instance, radiative exchange from a surface 1 to a surface 2 is given by:

$$q_{r12} = \varepsilon\sigma A_1 F_{12}(T_1^4 - T_2^4) \qquad (2)$$

where ε = emissivity, σ = Stefan-Boltzmann constant (= 5.67 10^{-8} W/m^{2}K)
A_1 = area of emitting surface, F_{12} = view factor appropriate to geometry
T_1, T_2 = absolute temperature (K) of surfaces, q_{r12} = radiative transfer from 1 to 2

The emissivity of most building materials including concrete is around 0.9 [2]. The view, or configuration, factor is a the fraction of the radiation leaving surface 1 which is intercepted by surface 2, and is dictated by the geometry.

Long parallel surfaces will have a view factor of one, whilst surfaces at right angles with a common edge (a ceiling and wall) will have a factor calculated from a knowledge of their respective dimensions.

Conduction

This occurs within the slab, as thermal energy from the surface layer, transferred by convection and radiation, is transmitted through the material (or vice-versa). It occurs in a liquid or solid electrical non-conductor by means of lattice vibrations at an atomic level, with lattice waves moving through the material in what amounts to a diffusion process. The rate equation which describes conduction is an application of Fourier's Law, with the constant of proportionality being the thermal conductivity k (W/mK). For the simplest case of one dimensional conduction, for instance perpendicular to the surface of a concrete slab (which may be defined as the x direction), the Fourier equation can be written:

$$q_x'' = (k/L)\Delta T \qquad (3)$$

where k = thermal conductivity (W/mK), L = thickness of slab
ΔT= temperature differential across the slab

Thermal Comfort and FES

Thermal comfort in an FES building is particularly influenced by the mean radiant temperature (the area-weighted average of the temperatures of surrounding surfaces), and dry resultant temperature (DRT) is often used as an indication of perceived temperature. At low air speeds:

$$DRT = \frac{\text{Air temperature} + \text{Mean radiant Temperature}}{2}$$

It can be seen that a cool surface, for example an exposed concrete soffit, decreases the DRT, and could mean that thermal comfort is maintained for higher air temperatures.

Implications for FES

To allow thermal interaction between the slabs and occupied spaces, they must be accessible to the air volume to enable convective heat transfer, and be 'visible' to other room surfaces and the occupants for radiative transfer. In other words, there should be good thermal linkage. It can be seen from the equations that heat transfer, and hence the effectiveness of FES, is greatest if a) the temperature differential between the concrete and air and between other room surfaces is maximized, and b) the surface area of concrete is maximized. The equations shown describe the steady-state condition, whereas FES is essentially dynamic, and requires more complex analysis, for which dynamic thermal simulation is required.

Design Approaches

Passive FES concrete surfaces are exposed to allow convective and radiative heat exchange with the air volume and surrounding surfaces respectively. Natural convection and thermal radiation play a roughly equal part. The dry resultant temperature is often below ambient because of the cool radiant temperature of the concrete surface.

Night cooling is by opening windows or vents, possibly using stack effect, or a conventional ducted mechanical system could be installed. A passive FES building can contribute a cooling effect of 15-20W/m^2, sufficient to counteract the effects of computers and printers at average usage densities [3].

Active FES achieves enhanced heat transfer by forced ventilation through a hollow core slab or across the slab surface by means of a plenum (for example, beneath a raised floor). The convective heat transfer coefficient is increased substantially, particularly if a turbulent flow regime is created at the slab surface.

A cooling effect of 25-35W/m^2 is often achieved, allowing comfort to be maintained at higher levels of internal heat gain without resorting to air conditioning. Active systems are more controllable than the passive approach, enabling ventilation rates to be changed in response to cooling demand, and in some cases allowing thermal linkage with the slab to be varied.

DESIGN FOR FES

As a heavyweight material, concrete offers substantial capacity for fabric energy storage (FES). Exposure of columns, beams, walls, slabs and soffits is an effective means by which the building structure can contribute to the 'thermal flywheel' effect. In theory the more mass exposed to the internal air volume, the greater FES attainable. However, designers should be aware that their efforts to utilise FES may be best aimed where there is the greatest scope. To attain maximum lettable area, commercial buildings now use very efficient structural designs with generally fewer structural elements. This means that the building frame or superstructure also offers little scope for FES [4]. Furthermore, much of the building skeleton along with ducting, trunking and other services, floor systems or dry lining is traditionally concealed behind suspended ceilings. Hence frame contractors may not be accustomed to producing visually acceptable concrete frame elements on a regular basis, although there is a developing capability in this area as more FES buildings are constructed [5,6].

Structural and Architectural Design

The main factor which influences how great an area of concrete can be exposed is the area of glazing required to satisfy daylighting requirements, but the nature of the facade is also important. Coordination of perimeter servicing or trunking and the position of insulation needs to be considered at an early stage. Concrete panels, masonry or blockwork acting as the inner structural or non-structural leaf of a cavity wall construction could be used to maximise FES (and vertical surfaces have a higher convective heat transfer coefficient). Panels could be cast in-situ using vertical formwork or horizontally on the ground slab and then lifted up into place (tilt-up construction), or the panels could be precast off site under factory conditions. Although there is evidence that concrete panels can be competitive with lightweight metal-intensive solutions, the latter tend to dominate the UK commercial property market [7]. However, the long-term benefits of increased FES and resultant energy savings for the occupier, should be analysed in a whole life costing exercise.

Frames offers little potential for FES, but may be worth adapting at the design stage (while still conforming to BS8110) to get some benefit. However, there are conflicting requirements of increasing surface area for FES versus reducing element size for structural efficiency to consider. Columns can have section sizes increased, but this conflicts with occupiers' needs for greater floorspace. However, surface area can be increased by creating texture or reveals, or by adding concrete at column heads where it will not impact on floorspace. However, it is more likely that structural efficiency will be the deciding factor, and the challenge is therefore to optimise FES capacity in a structurally efficient frame. The recent Lloyds Register of Shipping building in London uses precast scalloped downstand beams partly to support the precast trough floor, and partly to maximise FES. Floor slabs can contribute to FES while still being structural efficient. The most popular option is to expose the undersides of slabs and use raised access floor systems on the floor above. This is an efficient method because the upper surface will almost always be obstructed by furniture, so offers less FES surface area, while the slab underside (soffit) can absorb heat gains from occupants and equipment below. One of the simplest means is to arch or shape the soffit architecturally. Troughed, ribbed in-situ slabs or double-tee precast floors may offer a significant increase in surface area. Waffle slabs with deep downstands increase surface area of the soffit quite dramatically.

Bespoke construction methods have been developed to satisfy other requirements such as servicing, daylighting, acoustics and aesthetics. These use both in-situ and precast concrete construction. There is a body of opinion that precast construction is suitable for regular structural grids, whereas in-situ concrete is suited to more bespoke schemes. However this is not always the case. For a regular, uniform grid there is potential to capitalise on such economies of scale, and bespoke or irregular frame buildings are somewhat less suitable unless the precast moulds can be constructed with a degree of tolerance or flexibility. Exposed precast concrete arched soffits in Sir Richard Rogers' Lloyds Register of Shipping building (in which each successive line of troughs cants through 1-1.5 degrees on plan) demonstrate that precast can be used for congested city sites with difficult geometry. The Inland Revenue Building in Nottingham is more typical of a precast concrete structure with a regular, gently curving design. However the PowerGen HQ building described later shows that in-situ concrete can produce high quality finishes on a regular grid for a naturally ventilated passive FES building.

Concrete Specification and Finishes

This section considers mainly the attainment of acceptable finishes for soffits, but is equally applicable to frame elements and wall panels. Soffits, i.e. the undersides of the floor slabs. will be visible above work spaces, and should thus satisfy both the architects' and occupants' aesthetic requirements. Two basic characteristics of the concrete should be addressed to achieve the desired quality; surface consistency and colour consistency.

To attain an acceptable surface consistency, the designer should include a clause in the concrete frame specification to state the number and size of acceptable imperfections such as hair line cracks, blowholes and other features per unit surface area. Surface smoothness is affected by the concrete mix specification and by the material of the formwork or formliner on or against which the concrete is placed. To achieve smooth, consistent surfaces on soffits it may be necessary to use bespoke high quality steel or glass-reinforced plastic (GRP) moulds or forms (see PowerGen case study).

The concrete mix should also consider any requirements for colouring. FES soffits in particular are most effective (for daylighting and lighting) if they have a white or pale coloured finish which helps to reflect or 'bounce' light onto workspaces. Most exposed soffits are painted white which is advantageous to the contractor or precaster which can focus on producing concrete with excellent surface consistency without concern about paler or darker areas of colour. The advantage to the owner is that a paint finish can be changed at any time, and maintained easily.

CASE STUDIES

The following two case studies have been selected to illustrate how FES principles and different methods of construction can be used. The PowerGen HQ building uses in-situ concrete and a passive design approach. The Elizabeth Fry building uses an active design approach with precast hollowcore floors.

1: PowerGen HQ, Coventry

Bennetts Associates used in-situ concrete exposed soffits for the PowerGen HQ in Coventry, which opened in January 1995 and won acclaim for its architecture and environmental design. The 12,700m^2 three storey offices for PowerGen, one of three electricity generating companies in England and Wales, demonstrate a willingness on the part of office occupiers to adopt FES systems. This was said to be a successful FES building because the exposed concrete satisfied 'functional, structural and aesthetic requirements in one feature' [8]. Designers were instructed to create a productive and enjoyable workspace which was also environmentally friendly [9].

The design of the workspace influenced the plan of the building, and on the design of the FES soffits. Long term productivity improvements were thought best achieved through an open plan design, with narrow (13m) floors, a 113m long atrium, and natural ventilation which could be controlled by the individual.

PowerGen uses natural ventilation enhanced by a stack effect through the open plan offices and the three storey atrium. This reduces the need for mechanical ventilation, and placing heat producing equipment such as photocopiers in localised 'hot spots' means mechanical plant is kept to a minimum. The exposed concrete frame and soffits contribute significantly to comfort conditions, and so the building was designed without air conditioning. The soffits incorporate uplighting, downlighting, and acoustic absorption on a lighting raft, the PA system, smoke and fire detectors, and also rebates for partition heads.

The structural bays for PowerGen were determined mainly by the administrative environment of the building. The structural engineers were charged with developing ideas for the 10.8m span concrete soffit units over the office workspaces. Precast concrete was considered, but because the units would have weighed 20 tonnes, and cost 10% more than an in-situ solution, the in-situ proposal was chosen.

The frame contractor advised on buildability to ensure that this sophisticated elliptical, profiled slab and soffit would be as fast on site as any other in-situ concrete frame. The whole structure was erected in 16 weeks over the winter. The exposed in-situ concrete soffits were formed using elliptical GRP moulds to give the underside of the slabs a smooth, even finish which was suitable for painting with normal emulsion paint (see Figure 1).

Despite the additional cost of manufacturing the moulds, they were durable and re-used several times. Quality control, regular slump tests and sand from the same source helped to ensure the concrete for the soffits was very consistent. The soffit was designed to achieve maximum crack widths of 0.1-0.2mm, so that normal emulsion paint would cover blemishes. The principal engineer commented that the colour of the concrete was satisfactory and the surface finish was very good quality, confirming his belief that there is no need to use precast concrete for fear of not achieving quality with in-situ concrete. [10]

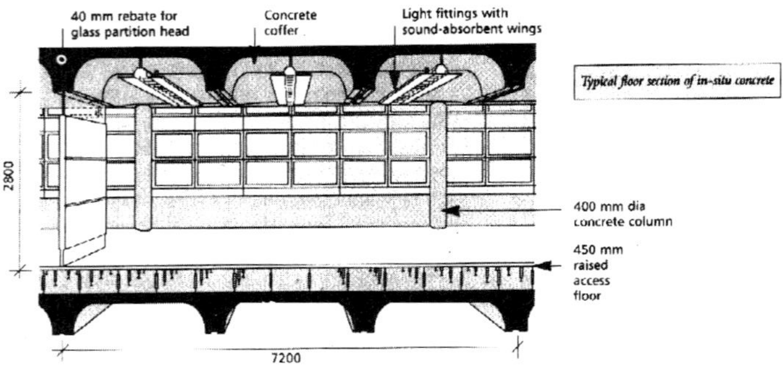

Figure 1 Concrete Design for FES at PowerGen HQ (Courtesy of RCC)

Although PowerGen is not complex, the design team were keen to ensure the environmental design would perform under peak summer temperatures. Environmental specialist EDSL was commissioned to model the predicted performance of the building based on known temperature swings for a hot spell [11]. The dynamic thermal simulation program calculated the peak internal temperatures as determined by external temperature, internal gains and the FES, and revealed that the in-situ concrete would perform very well over a long period of hot days provided it was cooled by ventilation 'purges' at night. Daylighting levels in the building were also simulated [12].

Table 1 Sample of predicted performance and actual performance (August 1995)

DAY	OUTSIDE AIR TEMPERATURE (°C)	INTERNAL AIR TEMPERATURE (°C)	ACTUAL REDUCTION (°C)	PREDICTED REDUCTION (°C)
1	29	26	3	3
2	28	26	2	2
3	28	26	2	2
4	28	26	2	2
5	29	24	5	4
6	27	25	2	1
7	31	26	5	4

Following occupation of PowerGen in January 1995, the building was monitored. As with the following case study, this process covered a mini-heatwave in summer 1995 in which outside air temperatures reached over 30°C. Data indicates that the internal temperature in Powergen reached a maximum of about 27°C despite outside air temperatures of 32°C, and inside temperatures were at least 3°C below outside temperatures during the warmest hours. Perhaps most indicatively, the inside temperatures remained within a 5°C range throughout this period, proving the early predictions of the FES 'lopping' peak temperatures correct.

Design Lessons for a Passive FES Building

- Exposed in-situ soffits can achieve a significant temperature lop and contribute to a well-tempered naturally ventilated environment even during a heatwave. A successful FES strategy must provide effective night-time ventilation routes through the building.

- Careful integration of structure, services, lighting, and acoustics in addition to buildability should be high on the agenda in design discussions as early as possible. The contractor or concretor should be involved in these discussions because such soffits are often part of a bespoke solution.

- Testing, modelling and mock-ups are vital to achieve the best possible ventilation and aesthetic effects. Quality of finish should also be displayed and agreed upon.

- PowerGen is a highly-regarded example of successful FES use in a contemporary office building. The high level of integration in the design team and the close involvement of the client have resulted in a very attractive working environment.

2: The Elizabeth Fry Building, University Of East Anglia (UEA)

The Elizabeth Fry Building (EFB), designed by architects John Miller and Partners and completed early 1995, is the latest in a series of low energy buildings commissioned by the expanding University of East Anglia. The University worked closely with the designers to complete a project which sets new standards for energy efficiency in this country. It has been chosen here because it demonstrates the effective use of precast hollowcore slab FES (TermoDeck). The building was monitored by one of the authors in the very hot summer of 1995. EFB provides 3250m^2 of varied accommodation in a shallow plan over four levels. The lower ground floor, accessible from the south of the building, houses four lecture theatres, washrooms and plant rooms. The ground and first floor consist of seminar rooms and offices on either side of a central longitudinal corridor, with the first floor also containing a common room in the curved bay over the main entrance.

The building is mechanically ventilated by air handling units in basement plant rooms, which supply air to the occupied spaces along corridor ducting with branches to the TermoDeck slabs. The central cores of each slab used for air supply are connected such that air travels through them in a serpentine fashion, entering the rooms from ceiling diffusers near the facade. Air is extracted via the light fittings to a central corridor plenum and exhausted at the basement after passing through a regenerative heat exchanger. As air travels along the cores of the slab, heat is transferred convectively between the air and the concrete surface (see Figure 2). Fans run during occupied hours and for night cooling if required. At night, after 22.00 hours, fans are switched on by the BMS to pass cool air through the slabs, picking up heat and cooling them to a minimum of 21°C. During a warm day, air is cooled by the slab before it enters the space, emerging at the diffusers up to 6°C cooler than ambient external temperature. This convective cooling is supplemented in about equal measure by radiant cooling from the exposed slab soffit. When heating is required, heat is transferred to the air stream from a heater battery supplied with hot water from three gas boilers in the plant room.

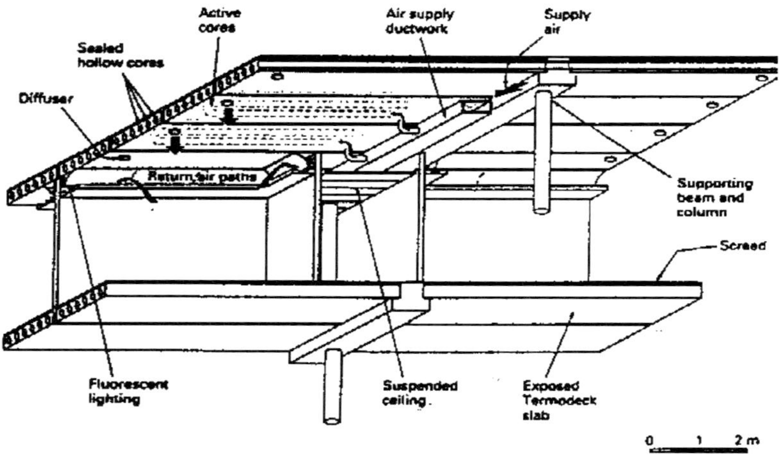

Figure 2 Integration of Termodeck and Ventilation System (CIBSE Journal, April 1995)

The design aim was to provide a very airtight, well insulated envelope incorporating exposed concrete mass in the form of hollowcore floor slabs. The building is supported by an in-situ concrete frame; each floor is formed from standard 1.2 m wide precast hollowcore slabs placed laterally to the floor plan, perpendicular to a longitudinal spine beam. The slab soffits are exposed to the room below, and painted white to improve reflectivity. Exterior walls with dense concrete blockwork were designed to provide additional thermal mass. Corridors and stairwells are of blockwork which again provides additional thermal mass (left unrendered on the corridor side), whilst partition walls are of conventional dry-lined construction.

Temperatures were logged over a three month period in three rooms of the building: a basement lecture theatre, a ground floor south-facing seminar room, and a first floor north-facing office. Sensors measured mean air and globe temperatures and the temperature of the air entering the room via the hollowcore slab. In the plant room, inlet/exhaust and supply/extract temperatures were recorded. The data were analysed as part of an MSc thesis at Cranfield University [13]. Figure 3 shows that internal conditions are substantially modified by the building, keeping conditions in the first floor seminar room comfortable over a series of very hot days. Air supplied to the slabs peaks at 28.1°C in the afternoon, and is cooled to 24.8°C by passage through the cores. The air temperature in the room at the same time is slightly higher, at 25.2°C, although the black globe temperature is lower (measured within a black spherical chamber), being 23.9°C. The globe temperature, which more closely approximates to perceived temperature is 6°C lower than the recorded ambient maximum on that day. The temperature of air entering the room from the slabs is only 0.4°C lower than the mean air temperature, and hence cool draughts are avoided. The timing of peak temperature in the building is notable. The outdoor temperature peaks at 30°C at around 15.00, whereas the indoor temperature exhibits a significant lag of 5 hours, reaching a maximum at 20.00, long after the occupants have departed for home.

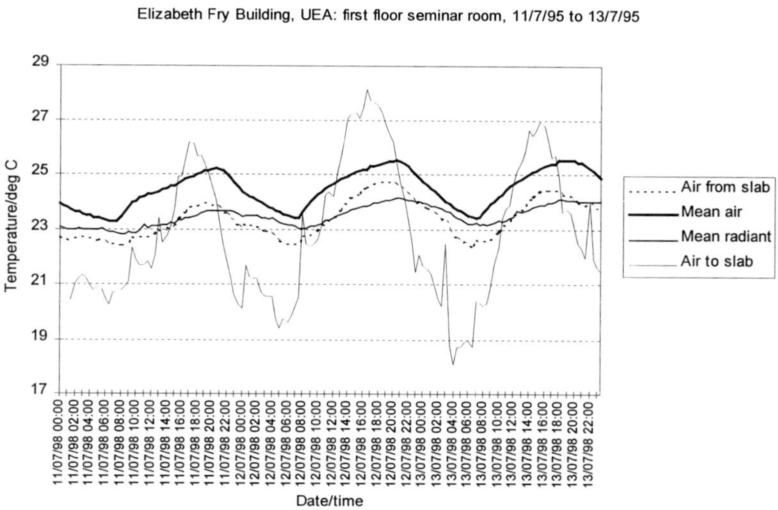

Figure 3 Typical Monitored Data for a Seminar Room at Elizabeth Fry

Design Lessons for Active FES Buildings

- Fan power should be kept low by designing a low pressure drop ventilation system. Use large cross-sectional area ducts and large radius turns with guide vanes if possible. Specific fan power in the order of 1W/l/s should be achievable. Air supply/extract routes should be considered early, so that sufficient space is allowed with minimum contortion of ducting. Low fan power maximizes the coefficient of performance (cooling effect divided by fan power used) and minimizes fan temperature pickup into the supply air stream (approximately 2°C at Elizabeth Fry).

- An exposed concrete structure can be aesthetically pleasing as well as functionally efficient, but careful attention must be paid to detailing and finishes.

- Site the air handling plant rooms in basements if possible where a lower average temperature can be maintained due to ground coupling and lack of solar gains even on a hot day.

- Air intakes should be located on the shady side of the building, not adjacent to potentially hot areas of tarmac such as car parks. Likewise, ducting should not be routed through hot solar heated roof spaces, and should be insulated with 25-50mm of insulation, to avoid excessive temperature rises en route to the slab during night cooling.

- Use of inlet/outlet heat exchange, such as the 85% efficient regenerative unit used at the EFB, allows useful heat gains to be stored during cooler months whilst full fresh air ventilation maintains excellent indoor air quality. Heating energy is thus much reduced.

The EFB has since been monitored by the Building Research Establishment, and has been assessed in the Building Services Journal's PROBE post-occupancy study [14].These studies confirm the high level of energy efficiency achieved, with electricity and gas consumption being 70 kWh/m^2/yr and 25 kWh/m^2/yr respectively. This is a total energy consumption of 95 kWh/m^2/yr, less than the latest DETR Good Practice figure of 112 kWh/m^2/yr for a naturally ventilated commercial building [15]. The inside air temperature has not exceeded 24.6°C since summer 1995, in spite of ambient temperatures above 29°C. In the occupancy survey, the building scored in the top 20% for criteria such as thermal comfort and indoor air quality. The Elizabeth Fry Building shows that hollowcore fabric energy storage using mechanical ventilation and heat exchange is an approach that, when implemented as part of an holistic low energy design, can achieve a high level of both thermal comfort and energy efficiency.

CONCLUSIONS AND RECOMMENDATIONS

This paper has examined the implications of basic heat transfer theory on design for FES in terms of exposed concrete surface area and finishes. Two case studies have illustrated the use of in-situ construction for a passive design approach, and the use of precast hollowcore for active FES. In-situ and precast concrete FES systems offer significant peak temperature reductions and satisfy demands for low operational and maintenance costs and thermal comfort provided they are well conceived, designed and constructed. Issues of servicing, surface finish, daylighting, lighting and acoustics and flexibility all have an effect on the feasibility of exposing concrete elements within the office environment. FES can be maximised by good design and construction, but also by providing clients and developers with evidence of the long-term thermal comfort improvement and operational energy savings of such a choice. If FES is to be used more widely in offices then it must exhibit clear benefits, including flexibility. In-situ concrete construction is most suitable for a passive approach to FES, and careful design and construction can produce aesthetically acceptable exposed surface finishes as demonstrated by the PowerGen HQ. Alternatively, precast hollowcore offers a greater degree of controllability when used in conjunction with an active FES design approach because heat transfer and ventilation can be varied. The Elizabeth Fry Building demonstrates the efficiency of the TermoDeck system when integrated into a well implemented low energy design, but warns that active FES solutions also require forethought to achieve an optimum mechanical layout. An established design protocol, or decision-making tool to predict internal conditions for different building types and design configurations should be made available to aid the design team in its choice of an appropriate FES system. This could be based on work carried out by the BRE on night ventilation [16].

REFERENCES

1. THE CHARTERED INSTITUTE OF BUILDING SERVICES ENGINEERS. Natural Ventilation in Non-Domestic Buildings, CIBSE Applications Manual AM10, CIBSE, London, 1997, p33.

2. THE CHARTERED INSTITUTE OF BUILDING SERVICES ENGINEERS. CIBSE Guide, Volume C, Reference Data , CIBSE, London, 1986, pp.C3-9.

3. AMATO, A., OGDEN, R.G., PLANK, R. Environmental Floor Systems, SCI/P/181, Steel Construction Institute, Ascot, Berkshire, 1997, pp.29.

4. GOODCHILD, C.H. Economic Concrete Frame Elements, British Cement Association, Crowthorne, Berkshire, 1996, 128pp.

5. BAICHE, B. Perception of Hybrid Concrete Construction Within the UK Construction Industry, Unpublished Report, School of Architecture, Oxford Brookes University, 1997, July, 15pp.

6. GOODCHILD, C.H. Hybrid Concrete Construction, British Cement Association, Crowthorne, Berkshire, 1995.

7. GLASS, J. Evaluation of Tilt-up Construction in Relation to Selected UK Building Types, School of Architecture, Oxford Brookes University, Unpublished Doctoral Thesis, 1997, 558pp.

8. O'NEILL, B.T., SHAW, G., & FLYNN, M. Project Profile PowerGen Headquarters, British Cement Association, Crowthorne, Berkshire, 1996, 13pp.

9. ALLISON, S., BENNETTS, R. ET AL The Choice of a New Generation, Architects Journal, 1995, vol.201, no.9, March 2, pp.43-55.

10. SHAW, G. A Consultants View of Concrete, Architecture Tomorrow, Architects Journal Centenary Conference, 1995, March 8, London.

11. ENVIRONMENTAL DESIGN SOLUTIONS LTD, TAS Building Simulation Software, EDSL, Milton Keynes, 2pp.

12. JACKAWAY, A. & GREENE, D. PowerGen in the Light of Day, Architects Journal, 1996, vol.203, no.10, March 14, pp.46-47.

13. KENDRICK, C. Use of Structural Components to Provide Night Cooling at a Building at the University of East Anglia, Department of Applied Energy, Cranfield University, Unpublished Masters Thesis, 1995

14. THE PROBE TEAM, PROBE 14 Elizabeth Fry Building, Building Services Journal, 1998, April, pp.37-42.

15. DEPARTMENT OF ENVIRONMENT TRANSPORT AND THE REGIONS. Energy Consumption Guide 19, Energy Use in Offices, 1998, p.20.

16. KOLOKOTRONI, M. Night Ventilation for Cooling Office Buildings, IP4/98, Building Research Establishment, Watford, Hertfordshire, 1998, March, 8pp.

STRUCTURAL BEHAVIOUR OF REINFORCED CONCRETE MASONRY BEAMS UNDER SHEAR STRESSES

S M Hilal

Cairo University

H H Bahnasawy

M A Khafaga

Housing and Building Research Centre

Egypt

ABSTRACT. The present investigation aims to study the behavior of the reinforced concrete masonry beams under shear stresses. In order to achieve the aim of study, nine simple span beams were manufactured and tested with two concentrated load. The cross section of each of the test beams consisted of U shape concrete blocks filled with concrete. All of these beams were under reinforced, ($\mu = (A_s / bd) = 0.36$ % and 1.03 %). The study aims to conclude the effects of shear span to depth ratio, (a / d ranged from 1.25 to 1.875), the amount of stirrups, (5 Ø 6 / m. and 10 Ø 6 / m.), the grade of filling concrete, (f_{cu} = 20, 30 and 45 N/mm²), and the presence of horizontal shear connectors between the blocks and the filling concrete on the type of failure. The results indicated that the shear failure occurs when decreasing the a / d ratio, the amount of stirrups and the grade of concrete filling and increasing the percentages of tensile reinforcement. Unexpectedly, increasing the grade of filling concrete caused increasing the recorded deflection of the test beam and no gain in using shear connectors between the blocks and the filling concrete. The present investigation illustrates that the width and the distribution of cracks in the test beams were correlated with the mortar joints positions and their number is very small when compared with the reinforced concrete beams.

Keywords: Masonry beams, Reinforcement, Shear stresses, Filling concrete, Shear connectors.

Professor Sameh M Hilal is a Professor in Structural Department, Faculty of Engineering, Cairo University, Egypt.

Professor Heba H Bahnasawy is a Researcher Professor in Strength of Materials Department, Housing and guiding Research Centre, Cairo, Egypt. Her main research interacts include properties of concrete especially in hot weather, quality control and quality assurance.

Mohamed A Khafaga is an Associate Researcher in Strength of Materials Department, Housing and guiding Research Centre, Cairo, Egypt. His main research interacts include reinforced concrete masonry and corrosion of steel in concrete.

INTRODUCTION

Many materials have been used for the construction of masonry, those locally available being most convenient. Whenever civilizations developed in river plains, the alluvial deposits were used to create brick architecture. In the Mesopotamian culture between the Tigris and Euphrates rivers, present-day mounds of crumbling soil in a treeless and stonewalls desert testify to the widespread use of sun-dried mud bricks in past buildings. The common masonry materials used today are made from stone, clay, calcium silicate, and concrete, and differ greatly in detail from those used in ancient masonry building [1].

In masonry construction, there are roof beams, floor beams, bond beams, and grade beams. Bond beams are usually located at the roof and floor levels and serve the dual functions of tying the building around the perimeter and acting as a chord member in floor or roof diaphragm action to transmit lateral loads to the shear walls. A lintel is a horizontal beam spanning over a door or a window opening and supports loads from the wall above.

All masonry beams must be reinforced because the limited tensile bond strength of masonry should not be relied upon to resist flexural tensile stresses from sustained gravity loads. For lightly loaded small spans, such as lintels over doorways, longitudinal reinforcement is often placed in bed joints. Reinforcement can also be placed in a grouted pocket or cavity formed by solid masonry units or in the grouted cores of hollow units, or in special shaped lintel units.

Neis and Loeffler in 1983 [2] studied the sheer failure of Reinforced concrete masonry beams, referred to as diagonal tension failure, which is a brittle mode of failure with very little deformation and therefore should be avoided.

Suter and Keller in 1976 [3] tested reinforced masonry beams. This study indicates that the shear behavior is similar to that of reinforced concrete beams. Masonry shear strength decreases with increasing shear span-to-depth ratio a/d, similar to reinforced concrete beams, increasing the reinforcement ratio leads to higher shear strengths of masonry beams.

Sutter, Keller and Fenton in 1984 [4] recorded test results for beams normally encountered in practice ($a/d \geq 3$), shear strength is not strongly correlated to the masonry compressive strength. For this reason, some codes specify constant shear stress values or relate the shear design stress to the tension design stress. They also proposed a shear strength design equation for masonry beams without web reinforcement in terms of a/d ratio and steel ratio. Lower a/d ratio and /or higher ~ values correspond with higher masonry shear strength.

Lee, Longworth, and Warwaruk in 1983 [5]conducted some tests, which indicated that web reinforcement in masonry beams prevents diagonal tension failure and allows the full flexural capacity to be developed.

Cyclic tests conducted by Lee and Neis in 1986 [6] on reinforced masonry beams with web reinforcement showed that the shear strengths of these beams were lower than similar reinforced concrete beams.

They attributed this to a number of possible effects such as poor bond between grout and units and the inherent weakness of the mortar joint planes that have very little mechanical interlock to increase shear resistance. Such results support the practice of neglecting the contribution of the masonry for cases where shear reinforcement is required.

Fereig in 1994 [7] tested 20 reinforced concrete masonry (RCM) beams with web reinforcement. The cracks in RCM beams propagated along the vertical and horizontal joints; thus; the nature of the crack propagation limits the effectiveness of vertical web reinforcement. Calculating the contribution of the vertical web reinforcement using the truss analogy approach similar to RC beams overestimates the contribution of vertical web reinforcement. The limits set by ACI 530/ASCE[(1988)codes] for the shear stress of beams with web reinforcement for working stress design may not provide an adequate safety margin.

The aim of this paper is to observe the behavior of the masonry beams under shear stresses.

EXPERIMENTAL DETAILS

The present investigation aims to study the behavior of simply supported reinforced concrete masonry beams under vertical concentrated loads. In order to achieve the aim of study, the schedule of the tested beams is shown in Figure l.

Test Beams

The test beams had a constant U shape cross section masonry concrete block filled with concrete. Figure 2 shows the shape and dimensions of the used concrete blocks.

Concrete Mixes

Three mixes of concrete are designed to achieve three characteristic strengths of 20,30,45 N/mm². The cement content was 250, 350, 450 kg /m³ for each mix.

Reinforcement

Plain round bars of 6, 10, 12 and 16 mm diameter have been used as tension reinforcement, compression reinforcement, and stirrups. The yield and ultimate strength were greater than 240 and 350 N/mm² and the elongation was greater than 20.

Masonry Blocks

The masonry blocks, which used in this study, were concrete blocks with a U shape. The average compressive strength of these blocks was 15 N/mm². The percentage of absorption was 5.5%. The modulus of elasticity of the blocks was 11000 N/mm².

TEST SETUP

Loading Apparatus

The loading system of the test beams is shown in Figure 3. The loading force was provided by a hydraulic jack 230 KN capacity. Loading was applied manually by a mechanical pump and load, strain, and deflection was measured automatically by a data acquisition system.

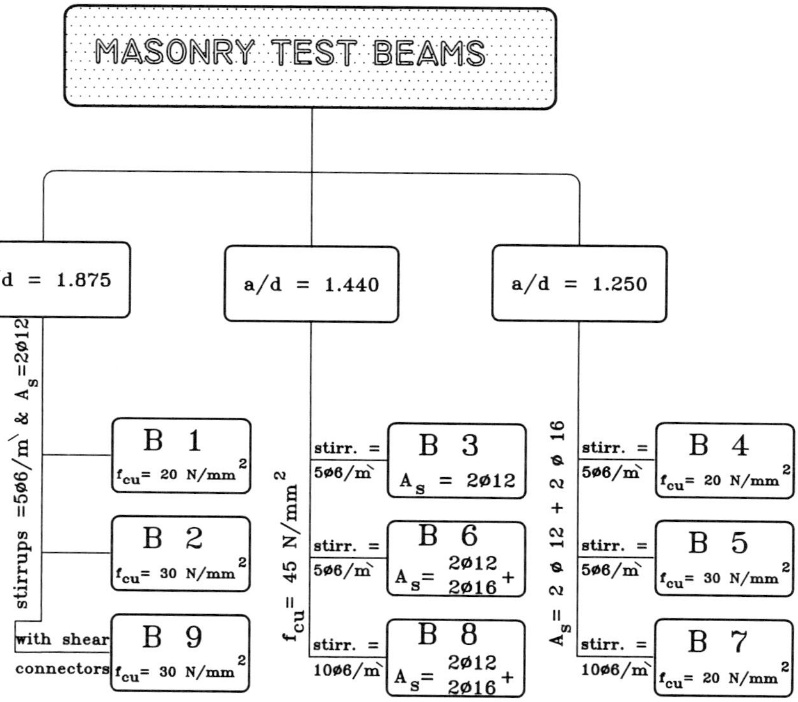

Figure 1 Schedule of the Test beams

Instrumentation

The load provided was monitored from the reading of the hydraulic jack gauge and checked from the readings of the data acquisition system. To control the flow of the pump of loading, a valve was conducted between the hydraulic jack and the pump.

The strain of the tensile reinforcement was measured by an electrical strain gauge 10 mm. length at the mid span of the beam. The strain of the stirrups was measured also by an electrical strain gauge at the third stirrup from the support.

The mean compressive and tensile strains were measured using a mechanical strain gauge of 200 mm. length, 5.0 mm. range and 0.01 mm. The deformations measured also at diagonal lines in of maximum shear region

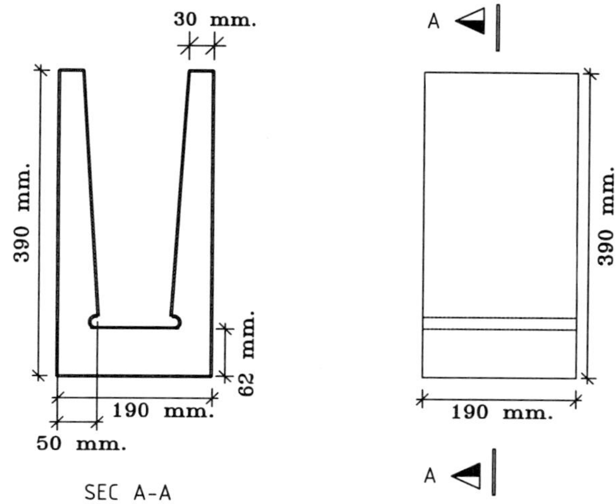

Figure 2 Details of the concrete block

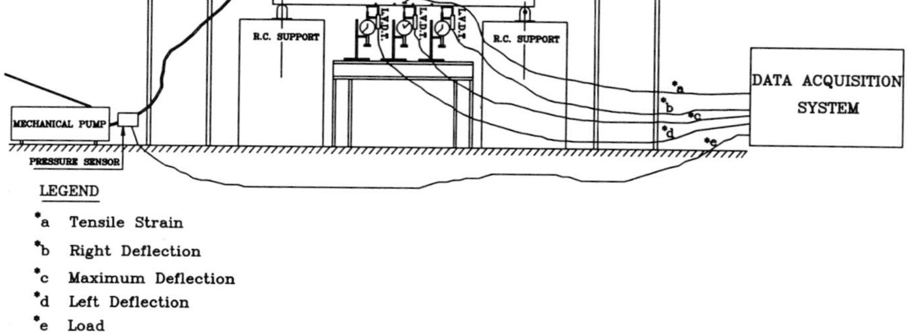

Figure 3 Test setup (loading system and instrumentation layout)

The beam deflections were measured using three dial gauges 50 mm. range and 0.01 mm. accuracy which shown in Figure 4 and checked from the data acquisition system.

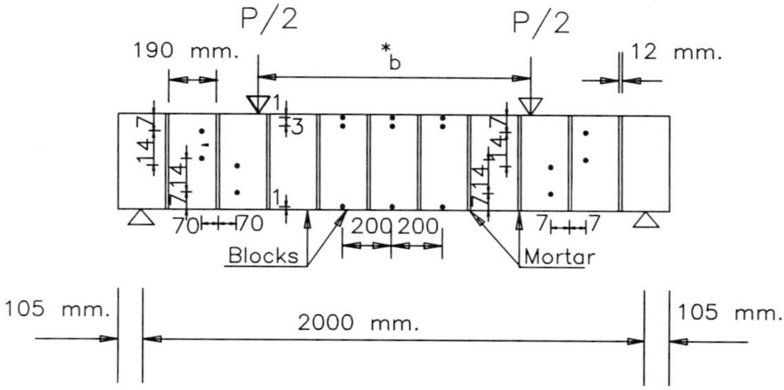

*

b= 800 mm. for beams, B1, B2 and B9

b= 1080 mm. for beams, B3, B6 and B8

b= 1200 mm. for beams, B4, B5 and B7

Figure 4 Test beams layout

Test Procedure

All beams were tested using an incremental loading procedure. Approximately 18-25 loading increments were used in the tests depending on the failure load for each beam.

RESULTS AND DISCUSSIONS

In order to investigate the factors affecting the behavior of reinforced concrete masonry beams under shear stresses, nine beams were tested. The span of all beams of this group was 2.00 m. The effects of characteristic strength of the filling concrete, percentages of the tensile reinforcement, shear span to depth ratio, amount of stirrups and presence of horizontal shear connectors between the blocks and the filling concrete on the mode of failure and the capacity of the test beams were studied. The strain distributions, the loaddeflection and the load-strains relationships and deflection lines were plotted for all of the test beams. Figure 5 illustrates the crack pattern for the beams of this group which failed under shear mode. The cracks were inclined at about 45° in the shear zones (between the support and the concentrated load). The cracks occurred in the blocks end the filling concrete together. This assures that there is full bond between the concrete blocks and the filling concrete. Figure 1 shows that these groups of beams were divided into 3 subgroups. The main variable for these beams was the shear span to depth ratio. Therefore, three ratios were used when testing these beams (a/d = 1.875, 1.44 and 1.25). The other factors, which were mentioned before, were also studied in these subgroups as followed.

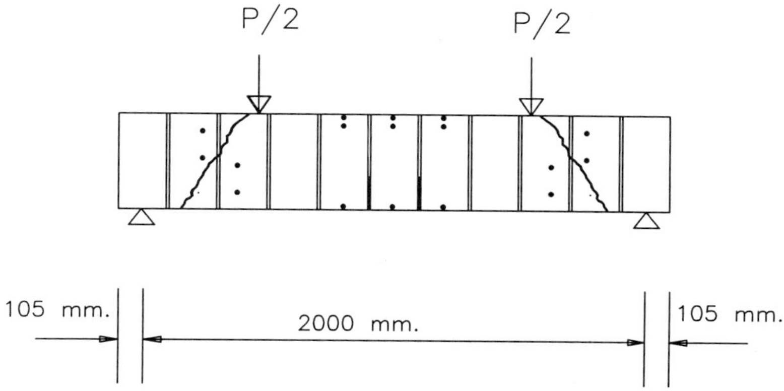

Figure 5 Crack patterns for beams which failed under shear

Beams with Shear Span to Depth Ratio Equal to 1.875

Effect of grade of filling concrete on the results for beams B1 and B2

Beams B1 and B2, having the same percentage of tensile reinforcement (p = 0.36 %ý), the same amount of stirrups (5 Ø 6 /m) and different grades of the filling concrete (20 N/mm² and 30 N/mm² respectively) were tested and the results of each of them were analyzed. Table 1 illustrates the yielding load and maximum deflection, maximum tensile strain in the tensile reinforcement and the maximum tensile strain in stirrups. One stirrup, which lies in the shear zone, was chosen (the second one from the support) and the strain of it was measured by an electrical strain gauge of 10 mm length.

Table 1 Results of tested beams

Beam No.	Filling Concrete compressive Strength (N/mm^2)	Actual yielding Loads (N/mm^2)	Diam. of Reinf. Used (mm)	Reinforcement Tensile Strains (KN)	Maximum Tensile Strain in Stirrups ($x10^{-3}$)	Maximum Central Deflection (mm)	Mode of Failure
B1	22.20	58.71	12	1.171	0.054	17.300	Flexural
B2	32.10	64.45	12	1.072	0.046	18.400	"
B3	46.80	91.32	12	1.700	0.372	19.300	"
B4	21.80	182.64	12	0.298	1.462	8.594	Shear
			16	0.176			
B5	32.90	218.62	12	0.412	1.449	7.905	"
			16	0.284			
B6	45.80	223.25	12	1.470	1.506	7.105	"
B7	22.20	198.88	12	0.323		7.810	"
			16	0.481			
B8	45.90	226.20	12	1.698	0.553	11.775	Flexural
B9	32.90	68.73	12	1.235	0.049	19.800	"

The following remarks can be deduced.

1. The two test beams failed under flexure. The tensile reinforcement reached yielding and the tensile strains in stirrups were much lower than the yielding value.

2. Yielding load for B 1 was smaller than that for B2.

3. Before yielding points, the recorded deflection for the two beams were approximately the same.

4. After yielding, the behavior of the two beams varied, the increase in the applied load for B2 was bigger than the increase in the applied load for B 1.

Effect of horizontal shear connectors on the behavior of the beams

For studying the effect of presence of horizontal shear connectors between the concrete blocks and the filling concrete, the results of tests of beams B9 and B2 were compared.

Table 1 illustrates the actual yielding load and the maximum central deflection and maximum tensile reinforcement strain respectively. It can be noted that the increase in the capacity of the beam was not big when using horizontal shear connectors. The recorded deflection values for B9 were smaller than the recorded values for B2 at the same applied load.

Beams with Shear Span to Depth Ratio Equal to 1.44

Beams B3, B6 and B8 were tested to study the effects of the percentages of tensile reinforcement and the amount of stirrups on the mode of failure and the capacity of each of those beams.

Effect of tensile reinforcement percentages on the results for beams B3 and B6

To investigate the effect of tensile reinforcement percentages on the mode of failure, beams B3 and B6 were tested and the results were compared and analyzed. These two beams have the same grade of the filling concrete ($f_{cu} = 45$ N/mm^2), the same amount of stirrup (5 Ø 6 /m) and different tensile reinforcement percentages ($\mu = 0.36$ % and 1.03 % respectively). From Table 1, which illustrates the yielding load and the maximum central deflection, maximum tensile strain in the tensile reinforcement and the maximum tensile strain in stirrups, the following remarks can be deduced.

1. Before yielding point for B3, the recorded values of deflection for B6 were always smaller than those for B3. It means that the increase in the tensile reinforcement percentages decrease the deflection of the test beam.

2. The sudden failure, which occurs for beam B6 assures that the mode of failure of this beam, was shear failure. It means that the mode of failure may be varied when increasing the tensile reinforcement percentages.

3. The tensile reinforcement for B6 did not reach at yield, i.e. the failure did not occur due to flexural stresses.

4. It is clear that the recorded strain in stirrups for B3 was very small and for B6 it reached at yielding value.

Effect of the amount of stirrups on the results for beams B6 and B8

To investigate the effect of the amount of stirrups on the mode of failure, beams B6 and B8 were tested and the results were compared and analyzed. These beams have the same grade of filling concrete (f_{cu} = 45 N/mm²), the same percentages of tensile reinforcement (μ = 1.03 %) and different amount of stirrups (5 Ø 6 /m and 10 Ø 6 /m respectively. Table 1, which illustrates the actual yielding load, maximum central deflection and the maximum tensile strain, which, indicate the followings:

1. The increase in the amount of stirrups increased the ultimate load of the test beam, the ultimate load for B6 and for B8.

2. From this Table it is clear that B8 which had the biggest amount of stirrups failed in flexural mode.

3. The recorded deflection for the two beams (B6 andB8) can tee considered the same before yielding point.

Beams with Shear Span to Depth Ratio Equal to 1.25

Beams B4, B5 and B7 were tested to study the effect of the characteristic strength of the filling concrete and the amount of stirrups on their behavior. Because of the shear span to depth ratio was very small, it can be expected the mode of failure will be shear failure for all of these beams.

Effect of grade of filling concrete on the results for beams B4 and B5

Beams B4 and B5 which were mentioned in Table 1 have the same percentages of tensile reinforcement (μ = 1.03 %), the same amount of stirrups (5 Ø 6 /m) and different grades of the filling concrete (20 N/mm² and 30 N/mm² respectively) were tested and the results of each of them were analyzed. Referring to Table 1 which illustrates the actual yielding load, maximum central deflection and load-maximum tensile strain in reinforcement, it can be deduced that:

1. Sudden shear failure occurred for the two test beams.

2. The increase in the grade of filling concrete leads to increase in the capacity of the beams which fail with shear failure mode such as the reinforced concrete beams.

3. The difference between the recorded deflections for the two beams at the same loading stage was very small tell about 140 KN applied load. After this load B4 deflected with bigger values and the difference between the curves of the two beams increased.

4. The recorded strain values were very small and the reinforcement was still et the elastic stage and shown curves can be considered straight lines.

Effect of the amount of stirrups on the results for beams B4 and B7

To investigate the effect of the amount of stirrups on the capacity of the masonry concrete beams, B4 and B7 which were mentioned in Table 1 and had the same grade of filling concrete (f_{cu} = 20 N/mm²), the same percentages of tensile reinforcement (μ = 1.03 %) and different amount of stirrups (5 Ø 6 /m and 10 Ø 6 /m respectively) were tested and the results were analyzed and compared.

From Table 1, the following remarks can be deduced.

1. Sudden shear failure occurred for the two test beams.

2. The increase in the amount of stirrups reads to increase beam.

3. The two beams had the same deflections tell about 140 KN applied load. After this load the difference between the two beams increased.

4. The recorded strain values were very small and the reinforcement still in the elastic stage and the curves can be considered straight lines.

CONCLUSIONS

1. There is a full bond between the blocks and the filling concrete, i.e., the block function integrally with the filling concrete in the compression zone.

1. Under shear failure, the ultimate loads of the test beams are approximately equal to ultimate load of the reinforced concrete beams.

2. The increase in the tensile reinforcement percentage leads to a proportional increase in the capacity of the beams.

3. The use of filling concrete with high characteristic strength leads to increasing in the deflection values and increase the capacity of the beams under shear failure.

5. The shear span to depth ratio effect on the beam failure mode (shear failure or flexural failure). The mode of failure is affected also by tensile reinforcement percentages and the amount of stirrups.

6. The presence of horizontal shear connectors between the blocks and the filling concrete does not affect on the results, hence it is not structurally or functionally needed.

REFERENCES

1. DRYSDALE, R G; HAMID, A A, and BAKER, L R. Masonry Structures Behavior and Design, Englewood Cliffs, New Jersey 07632, 1993.

2. NEIS, V V, and LOEFFLER, R J. Results of Ultimate Flexural and Shear Tests on Reinforced Masonry Beams, in Proceedings of the Third Canadian Masonry Symposium. University of Alberta, Edmonton, 1983, p.13/1-13/16.

3. SUTER, G T, and KELLER, H. Shear Strength of Reinforced Masonry Beams and Canadian Code Implications, in Proceedings of the First Canadian Masonry Symposium, Calgary, Alberta, 1976, p. 149- 160.

4. SUTER, G T, KELLER, H, and FENTON, G A. Summary of a Decade of Reinforced Masonry Research at Carleton University, Department of Civil Engineering, Carleton University, Ottawa, 1984.

5. LEE, R; LONGWORTH, J, and WARWARUK, J. Behavior of Restrained Masonry Beams, in proceedings of the Third Canadian Masonry Symposium, Edmonton, Alberta, 1983, p. 37/1 - 37/16.

6. LI, D, and NEIS,V V. Performance of Reinforced Masonry Beams Subjected to Reversed Cyclic Loadings, in Proceedings of The Fourth Candian Masonry Symposium, Vol. 1, Predricton, N.B, June 1986, p. 351-365.

6. FEREIG, S M. Shear Strength of Reinforced Concrete Masonry Beams With Web Reinforcement, y ACI Structural Journal, Vol. 91, No. 4, Jul-Aug.1994, p. 415-422.

7. KHAFAGA, M A M. structural behavior of reinforced concrete masonry beams under flexural and shear stresses M. Sc. Thesis, Faculty of Engineering, Cairo University, 1997.

CONCRETE INFILLED COLD FORMED STEEL RECTANGULAR BEAMS

A Swamidurai

L Ramajeyam

Anna University

India

ABSTRACT. Cold formed steel concrete composite sections are widely used in earthquake prone zones because of their high strength and ductility. These sections can be constructed easily by using prefabrication techniques and hence they can be used in multistory buildings and offshore structures. In this paper the theoretical and experimental behavior of cold formed steel composite beams subjected to gradually applied static load has been examined. Four groups of beam specimens of different wall thickness 1.2 mm and 2 mm are tested under simply supported condition. Welded mesh is introduced inside the shell to take care of the interslip forces and their effect is also studied and reported. Observations showed that due to confinement of concrete, strength and ductility have been improved. Debonding has been delayed because of the presence of weld mesh

Keywords: Cold formed steel, Weldmesh, Composite sections, Confinement, Static load, Strength, Ductility, Debonding.

Dr A Swamidurai is Professor of Structural Engineering, Anna University, Chennai, India. He specialises in cold formed steel concrete composite structural components with special emphasis on prefabricated structures. Professor Swamidurai has published widely and serves in many Technical Committees.

Er L Ramajeyam is a Research Scholar in Structural Engineering Division, Anna University, Chennai, India. He specialises on the composite structures and in pre-fabrication technique for multistory buildings.

INTRODUCTION

Since few decades, thin walled cold formed steel structural members have been widely used in the construction of different types of structural systems. The cast-in-situ beams require providing and removal of form work and shoring which constitutes a considerable part of their cost. However this construction cost can be significantly reduced if the conventional reinforcing is replaced by a thin walled, cold formed steel, functioning as form work before concrete attains its strength and as reinforcing steel after concrete attains its strength. Such beams may be used in the multistory buildings, bridges, cast-in-situ pile foundations etc.

By filling thin walled hollow rectangular sections with concrete, the failure of steel shell due to buckling will be delayed and the ductility of the concrete core will be improved as a result of confinement. These members have high strength, high ductility, easy construction and overall economy compared to the similar reinforced concrete members. In addition, resistance to fire, earthquake, impact are also improved.

THEORY OF CONFINEMENT

Concrete is a brittle material. Because of the limitation on the ultimate compressive strain that is feasible for normal concretes, varying between 0.003 to 0.0035, the failure deformation, namely ultimate deflection and curvature at failure are rather limited. Tests on cylinders and prisms with confining reinforcement have shown that the failure strain can be considerably increased. All the theories available to date assume that the core concrete exerts a `pressure loading' on the outer layer. Confinement really means `holding up to these pressures developed in the core by the outer shell'.

The improvement of failure strain in concrete by using cold formed steel box section has been reported [3]. Connection of the outer box shell by use of welded wire fabric and ensuring better shear transfer should logically improve the behavior and this aspect is being investigated on simple structural elements for exact quantification.

THEORETICAL ANALYSIS

The moments at cracking, yielding and ultimate and the corresponding curvatures have been estimated using a computer program. The strain and force distribution are shown in Figure 1. Theoretical M-ϕ curve for the beams are obtained and these have been used to compute the theoretical values of deflection corresponding to the loads using the moment area theorem. Idealised theoretical M-ϕ curve is shown in Figure.2.

EXPERIMENTAL DETAILS

Materials

Ordinary Portland cement of 43 grade was used to prepare concrete mix. The water cement ratio was maintained at 0.40(by unit). The above concrete mix had a slump in excess of 250mm. Cold formed steel sheets were used as the test specimen. Values of Young`s modulus, yield stress, ultimate stress and percentage elongation were obtained from test results.

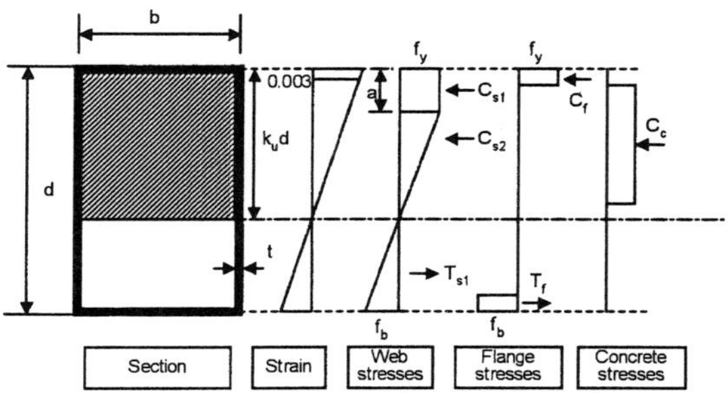

Figure 1 Strain and force distribution

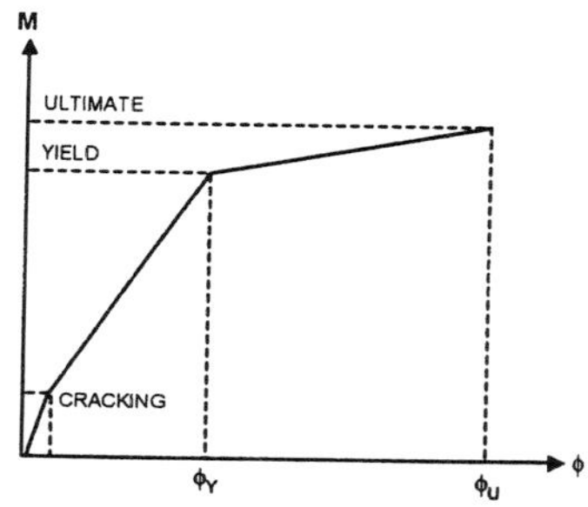

Figure 2 Idealised theoretical M-ϕ model

Fabrication of Test Specimens

The cold formed steel sheets were cut and folded to the required dimensions to form channel sections. Similarly the weldmesh were also folded. The weld meshes were tack welded to the steel tubes at every 150mm to secure them firmly to the steel sheets. Two folded sections were assembled together to form a box section of the required dimension with continuously welded longitudinal joints. Control concrete cubes were cast in standard steel moulds for each batch of mixing. Dimensions and details of test beams are shown in Table.1.

Test Setup

The beams specimen was tested on Simply Supported Condition. Three deflectometers were fixed one at the mid span and the other two at the load points enabling to measure the deflections at these points with periodic load increment. To measure the strain Demec studs were fixed at top and bottom of the beam in the pure flexure region and also in the mid span section of the beam.

Table 1 Details of Test Beams

SPECIMEN NO.	DESCRIPTION OF SPECIMEN	THICKNESS OF SHELL (mm)	DIMENSIONS L x B x D mm
E_1,E_2,E_3 & E_4	Cold formed steel concrete rectangular beam sections infilled with concrete	1.2	3300x80x120
F_1,F_2,F_3 & F_4	Cold formed steel concrete rectangular beam sections infilled with concrete	2	3300x80x120
G_1,G_2,G_3 & G_4	Cold formed steel concrete rectangular beam sections stiffened with 25 mmx 25 mm, size 3.2 mm dia weldmesh infilled with concrete	2	3300x80x120
H_1,H_2,H_3 & H_4	Cold formed steel concrete rectangular beam sections stiffened with 16 mmx 16 mm, size 1.51 mm dia weldmesh infilled with concrete	2	3300x80x120

TEST RESULTS AND DISCUSSION

A typical beam specimen, its test setup before loading, deflected shape during loading, deflected shape after test are shown in Figure 3.

Typical beam specimen

Beam after testing

Beam before loading

Beam during ultimate loading

Figure 3 Beam specimens before and after loading

Cracking of Core Concrete

The concrete core is completely surrounded by the cold formed steel plate and hence cracks in the concrete if any are not visible even after the core has cracked. Theoretical values of cracking moments and cracking curvature have been worked out by assuming the transformed section properties for the beam and modulus of rupture of concrete f_t as 0.7 f_{ck}.

Cracking of core concrete can lead to possible debonding between steel and concrete in the tension zone and if this happens a clear kink is to be seen in the M - ϕ curve. The experimental values of M - ϕ curve for beams E shows a clear kink approximately at the cracking moments level as seen from Figure 4. Provision of welded mesh wire fabric in beams G&H arrests the tendency of debonding and kink in M - ϕ curve have shifted significantly upwards.

The curvatures at 3.0 kN.M level, for beams, F, G and H are $9\text{x}10^{-3}$ rad/m , $7\text{x}10^{-3}$ rad/m and $5\text{x}10^{-3}$ rad/m, respectively. The extrapolated uncracked behavior appears to hold good up to 3kNm in beams F. while beams F have a slope of $3.5\text{x}10^{-3}$ rad/m per kNm between 1 and 3 kNm levels, the G and H beams have an average slope of $2\text{x}10^{-3}$ rad/m per kNm between 1 and 2 kNm, clearly indicated better structural behavior of the later group of beams ie (Group G & H).

Load Deflection behavior

All the beams showed failure deflection in excess of 80mm, corresponding to span to deflection ratio being 37.5, which shows a highly ductile behavior of the beam Figure 5. The failure rotation at the support using plastic analysis is (80/1500)rad i.e. 3.05°. Normal R.C. structural elements show only 1° to 2° rotation at support at the time of failure. Assuming a load factor of 1.50 and material safety factors (γ_m) of 1.15 and 1.5, the deflection at working stage is around 15mm for all beams. This corresponds to the (l/δ) ratio of 200 which is acceptable. Hence the overall behavior of all test beams can be considered as acceptable.

The ultimate load has increased significantly by about 45% even though the increase in the tensile steel area is only about 10%. This may be due to effective confinement provided by the outer shell when reinforced with welded wire mesh.

Validation of Test Results

Table 2. shows the theoretical ultimate moments computed using the values based on the beam model used in the report and the experimental values obtained from tests for all the specimens. The theoretical ultimate moments obtained are very close to the experimental values. The ratio of ultimate moments Mu (test/predict) ranges from 0.94 to 1.30 with the mean values of 1.09 which validates the test results.

Table 2 Comparison of theoretical and experimental ultimate moment

SPECIMEN GROUP	THEORETICAL M_u (kNm)	EXPERIMENTAL M_u (kNm)	RATIO OF (TEST/PREDICT)
E	10.37	9.75	0.94
F	12.32	12.75	1.03
G	15.89	17.00	1.07
H	15.10	19.50	1.30

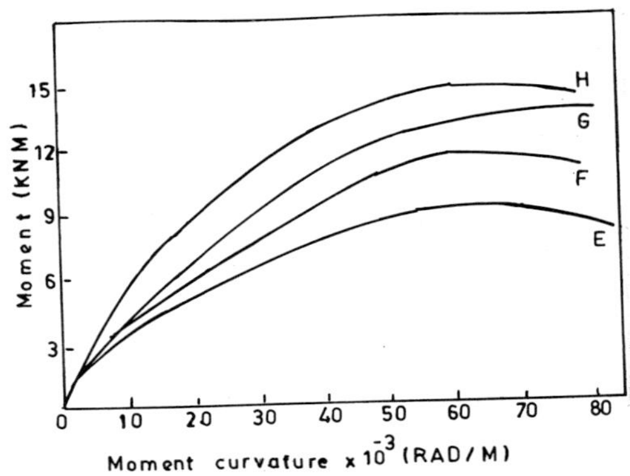

Figure 4 Moment curvature

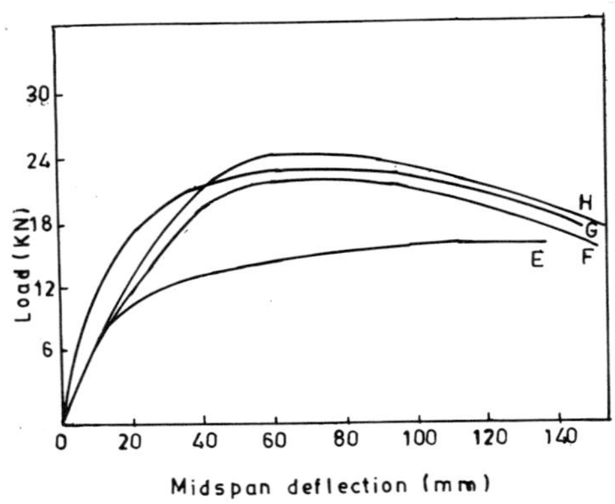

Figure 5 Load deflection curve

CONCLUSIONS

1. Provision of welded mesh wire fabrics as stiffener, arrests the tendency for debonding and the kink in experimental M - ϕ diagram shifts upwards as expected.

2. While the beams without weldmesh (Group F) has suffered an average curvature of 3.5 x 10^{-3} rad/m per kNm between 1 and 3 kNm levels, the beams with weldmesh stiffener (Group G) has curvature of 2 x 10^{-3} rad/m per kNm. This increased rigidity about 1.75 times indicates clearly the better structural behavior of beam (Group G&H) which has the weldmesh stiffener.

3. Stiffness is enhanced by 1/3 with a 50% increase in area of steel. As the area of steel increases, stiffness also rises, which is a similar phenomena to reinforced concrete beams.

4. The concrete prevents inward local buckling of the steel shell leading to use of the full effect of the cold formed steel shell up to failure.

REFERENCES

1. NGUYAN, R P. Thin walled cold formed steel composite beams. Journal of Structural Engineering, Vol.117, No.10, Oct.1991, pp.77-86.

2. LAKSHMANAN, N., PARAMESHWARAN, V S, KRISHNAMOORTHY, R K and BALASUBRAMANIAN, K. Ductility of flexural members reinforced symmetrically on tension and compression faces. Indian Concrete Journal, August 1991, pp.84-93.

3. ZHANG, J Q and BRAHMACHARI, K. Flexural behavior of rectangular tubular sections filled with fibrous high strength concrete. Tubular Structures VI, Grundy, Holgate and Wong (Eds), 1994, pp.171-186.

4. PARK, R. and PAULAY, T. Reinforced Concrete Structures. John Wiley and Sons Publication, New York, 1975.

5. PURUSHOTHAMAN, P. Reinforced Concrete Structural Elements. Behavior, Analysis and Design, Tata Mc Graw Hill Company Publication, New Delhi, 1984.

6. IYENGAR, K T S, NAGI REDDY, K and PRAKASH DESAI. Confined concrete its properties and applications, Proceedings, Forty Second Annual Research Session, Madras, 1970, pp.162-176.

ANALYSIS AND SPOT-TEST ON A SPECIAL COMBINED STRUCTURE INCORPORATING BEAMS, PLATES AND NET SHELLS

Gui-Ying Wu **Zhu Li**

Taiyuan University of Technology

Yao Li **Zhi-Ping Kang**

Twelfth Engineering of Railway Ministry

China

ABSTRACT. This paper presents the analysis of an upper large space special combined structure by finite element method, and the serial on-the-spot tests results of this structure in which the tension of nonagglomerating prestressing component was controlled by computer. The test results are identical with static calculations, which directly verify the correctness of the calculation and design. On-the-spot tests and control guarantee that the relative error rate between the final result of tensile force and designs is less than 5%.

Keywords: Special and complicate structure, Nonagglomerating prestressing tension, Finite element method

Professor Gui-Ying Wu works in Taiyuan University of Technology, China. She is interested in the elastic-plastic dynamics, anti-seismic structure of building and experiment mechanics.

Mr Yao Li is a senior engineer of Twelfth Engineering of Railway Ministry, China. His main research interests include the elastic-plastic dynamics and anti-seismic structure of building.

Mr Zhi-Ping Kang is an engineer of Twelfth Engineering of Railway Ministry, China. His main research interests include the elastic-plastic dynamics and anti-seismic structure of building.

Professor Zhu Li works in Taiyuan University of Technology, China. His main research interests include the elastic-plastic dynamics, experiment mechanics and anti-seismic structure of building.

INTRODUCTION

Finite Element Method has been wildly used in the analysis of special and complicated structures. On-the-spot tests are used in construction of these structures, which is of an unique function in improving quality of engineering, avoiding accident of engineering and verifying the correctness of design.

Shanxi Daily Office Hall is located in the center of Taiyuan City and is shown in Figure 1(a). Its height is 14.1m, and the maximum diameter of the upper bowl shape structure is 34m. This building fully utilizes upper space. It is challenging in the analysis, design and control tensioning of the three circle beams.

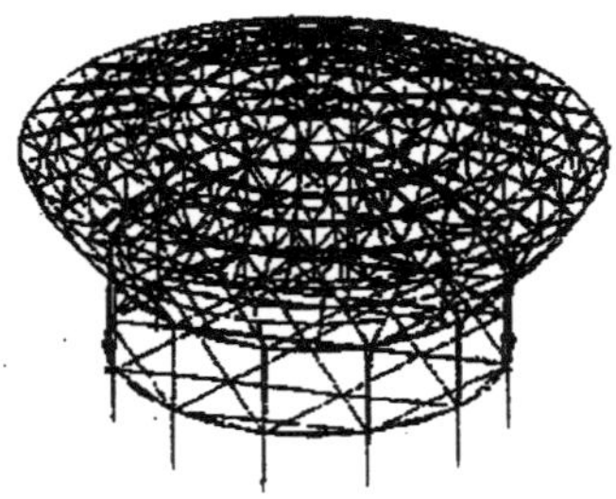

Figure 1 (a) The structure of Shanxi Daily Office Hall (b) The global computation model

The force sensors as made by the authors control the tension. The signals of the force sensors are processed on the spot by computer. In fully considering the adverse circumstance condition of the building site, the testing instrument is well deployed. This building has been erected and put into use.

STRUCTURAL ANALYSIS

Computation Model

In order to analyse this structure, it is necessary to select a computation model correctly. Considering complication of the structure, We select the global model. The final computation model refers Figure 1(b) and makes the hypothesis as following:

1. Taken the base as fix constraints
2. Taken the external load, snow load, active load etc. as volume force
3. Not considering the favorable function of the column and wall against earthquakes
4. Beams and floors act on together
5. The loads on the net shells are taken as single load acted on the other parts is uniform load.

Meshing

There are three types of elements, i.e. plate, beam, and truss element. The total element numbers are 1573, including 84 elements of plate, 360 elements of beam and 1129 elements of truss. The total node numbers are 472, and the total equation numbers are 1770.

Computation Results

The final computation results of Shanxi Daily Office Hall are as following:

1. The maximum response of earthquake at the bottom of column is 1.48 times static force, and at the other parts is 5-30% of static force. The response of earthquake becomes the smaller near the upper parts, which is according with the practice.
2. The first and the second vibration mode are all horizontal, which is dominating, refer to Figure 2.

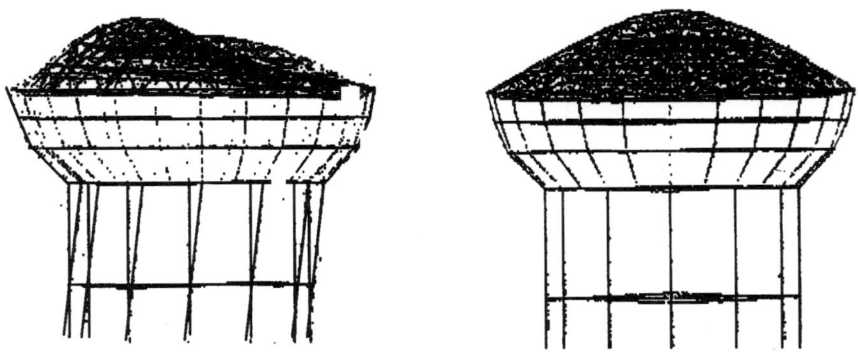

Figure 2 (a) The structure of Shanxi Daily Office Hall, (b) The global computation model

3. The period of free-vibration is high because of the cross sections of the building being large and stiffness of the column against earthquake being small.
4. The second vibration mode is dominating, which takes up 98% of the whole response, refer to Table 1.

Table 1 The coefficient table participated in vibration mode

MODE	X-DIRECTION	Y-DIRECTION	Z-DIRECTION
first vibration mode	-0.19e-1	-0.37e-2	-0.12e-6
second vibration mode	-0.37e-2	-0.19e-1	-0.43e-6
third vibration mode	-0.56e-1	-0.56e-1	-0.14e-2

INTERNAL FORCES AND DEFLECTIONS TEST ON THE SPOT

The internal forces of the first and second floor and beams as well as the net shells have been measured by strain gauge on the spot. These measurement points are selected according to structural symmetry and the results of computation. Total 94 strain patches are embedded in the structure. Testing time are selected at night as to acquire higher precision. The deflection of structure is measured by the deflection instrument, which are located on two temporary pillars that are erected on the spot. The sketch map refers Figure 3. The results of internal force and deflections measured are shown in Table 2.

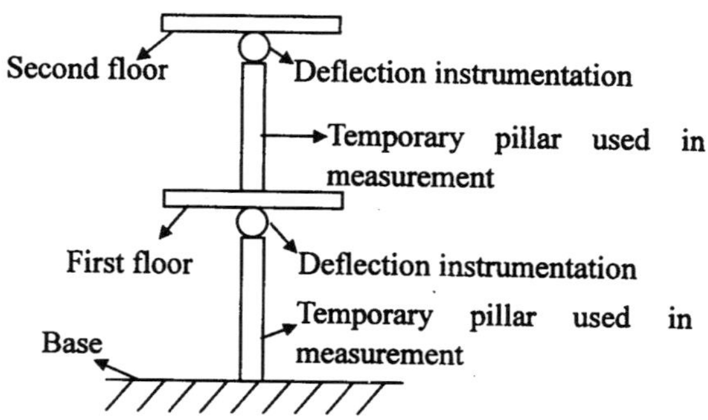

Figure 3 The sketch map of measurement points

Table 2 Comparison as measured and design results

	FIRST FLOOR		SECOND FLOOR		NET SHELLS	CIRCLE BEAMS
	Max. Mom (KNm)	Deflection (mm)	Max. Mom (KNm)	Deflection (mm)	Max Force (kN)	Max. Mom (kNm)
DESIGN	61200	29.0	58.69	33.0	128.7	61.33
ACTUAL	56900	31.8	56.20	31.5	123.3	59.30

* Mom = moment

TENSION CONTROL

The tension control of three prestressing circle beams is very important, for strength loss can not be afforded only by nonprestressing parts. On the other hand, if the tension is too high this will cause structure to be destroyed, so the precise tension must be reasonably designed.

The internal forces of the structure equal to the internal forces of prestressing parts plus the internal forces of nonprestressing parts, i.e.

$$\text{The design force} = \text{the prestressing forces} + \text{the nonprestressing forces} \qquad (1)$$

According to (l) and the computation results the design values of prestressing of three circle beams can be determined, which are respectively 80000N/member to upper circle beam, 50000N/member to second circle beam and 30000N/member to third circle beam. Final tensioning results must be within the limit of 5% of relative error rate between design and constraints make three circle beams influence each other in process of tensioning control. On considering of its large curvature effect and long tension length, we do not use conventional tensioning method. We have made the force sensors used in tensioning control and trial tensioning on the spot. In the process of trial tensioning, the duct loss values, and fixed loss values and the loss values with time of tension are all tested. All loss values calculated according the criterion is very different from the loss values that measured on the spot. According to the loss values measured on the spot and the formula (2), the tensioning control values can be confirmed.

$$\text{Tension control values} = \text{design values} + \text{all loss values} \qquad (2)$$

The circle beam should be firstly divided into several segments according to the symmetry of structure, the tension control values should be divided-into several grades in the process of tension and the tension must be operated at both end of duct at the same time. The final tensioning results refer to Table 3. Because we adopt force sensors and use computer to process the signal of sensors, the tension precision are guaranteed in the process of tensioning.

Table 3 The final tension results of circle beam

BEAM SERIAL NO.	SENSOR SERIAL NO.	STRENGTH, N/MEMBER		RELATIVE ERROR (%)
		Design (kN)	Measured (kN)	
upper	14	80	78	3
upper	12	80	76	5
upper	18	80	83	4
upper	16	80	84	5
upper	11	80	81	1
upper	19	80	82	3
upper	10	80	79	3
mid	7	50	52	3
mid	8	50	48	3
mid	9	50	48	4
mid	13	50	49	5
lower	4	30	31	3
lower	5	30	31	3
lower	1	30	29	3
lower	3	30	29	3

CONCLUSIONS

1 The results of internal forces and deflections measured on the spot have shown the correctness of the analysis results on the structure by Finite Element Method and the rationality of the design.

2. By adopting force sensors in combination with the computer tension method, the tensile forces of nonagglomerating prestressing component can be accurately controlled. Therefore, the design demand can be guaranteed very well.

3. Duct losses, fixed losses and the loss with time of tension must be tested on the spot in order to acquire the precise tensioning control values. .

4. The circle beam should be firstly divided into several segments according to the symmetry of structure. The tension control values should then be divided into several grades in the process of tensioning and the tensioning must be operated at both ends of the duct at the same time.

REFERENCES

1. GUI-YING WU ET AL. The Earthquake Response Analysis of Structure Constituted of Beam, Plate and Net Shell, Journal of Computation Mechanics, 1997, Vol. 14. 301-315.

2. ZONG-FANG YANG.. The Experimental Research of Tension Technology about Curve Prestress, Construction Technology, 1986, Vol. 8, 16-18

3. ZONG-FANG YANG. The Calculation of Fixing Loss of Curve Prestress, Building Structure , 1988, Vol. 8, 14-16

4. ZHU LI. The Structure Analysis and On-the-spot Experiment of Bowl Shape Structure, Building and Construction, 1998, Vol. 45, 18-22.

AESTHETIC PROPERTIES OF ARCHITECTONIC HIGH-PERFORMANCE SELF-LEVELLING CONCRETES

R Pleau J Henné

M Blais D Boulet

D St-Pierre

Laval University

Canada

ABSTRACT. This experimental study aims to compare the aesthetics properties of a self-levelling concrete with those of an ordinary concrete, both placed into vertical moulds having different shapes and made with different materials. Test results indicate that the use of a self-levelling mixture yields astonishing improvements as regards to the aesthetics properties of moulded surfaces. Self-levelling concrete provides smooth defect-free surfaces with very fme printed details, even for complex moulded shapes, and despite some particularly difficult casting conditions. The texture of the concrete surface was found to be strongly dependent of the material used for making the formwork, and a wide range of visual effects, ranging from mirror-like to broken stone appearance, were obtained. The improved aesthetics properties of self-levelling concrete are mainly due to the formation of a thin layer ($\approx$ 1 mm thick) of highly cohesive and workable cement paste near the mould wall. However, ASTM C 672 deicer salt scaling tests reveal that this thin layer could be vulnerable to frost action.

Keywords: Aesthetics, Self-levelling concrete, Rheology, Scaling, Microstructure.

Dr Richard Pleau (Civil Engineer) is Professor at the School of Architecture of Laval University. His research interests include fiber-reinforced cementitious composites. air entrainment, frost resistance, and high-performance concretes.

Mr Jerome Henne (Architect) is a Master Degree Student at the School of Architecture of Laval University. His thesis has for object the aesthetics properties of self-levelling concretes.

Dr Myriam Blais (Architect) is Professor at the School of Architecture of Laval University. She is particularly interested by the integration of the new construction technologies into the architectural design process.

Mr Daniel Boulet (Civil Engineer) is a Ph.D. Student of the Department of Civil Engineering at Laval University. His thesis has for object the micro-reinforced cementitious composites.

Mr Daniel St-Pierre (Architect) obtained his Bachelor Degree from the School of Architecture at Laval University in 1998.

INTRODUCTION

Aesthetics of structures is a major concern for architects. In the precast concrete industry, it is possible to achieve a satisfactory surface appearance because the placing conditions are favourable (concrete is poured into thin flat moulds and compacted using highly efficient vibrating tools) and it is possible to use special techniques to provide a wide range of surface colours and textures. This is hardly possible with cast-in-place concrete and, after demoulding, the concrete surface is rough and has many surface defects (voids, honeycombs, discoloration). As a consequence, concrete is generally considered as a cheap and ugly material, and architects are reluctant to leave it apparent. When possible, they cover it with stone or other cladding materials which are more attractive. Otherwise, they are resigned to accept the relatively poor appearance of concrete as a "fact of life".

The recent advent of self-levelling concretes could significantly contribute to improve the aesthetics of cast-in-place concrete surfaces. Their higher workability facilitates the compaction of concrete, which presumably reduces the presence of entrapped voids and other defects at the surface of the moulds. One can also expect that the use of a self-levelling mixture would not only reduce the number of defects, but could also modify the smoothness of the concrete surface. The objective of the test program was thus to investigate how the use of a self-levelling concrete can improve the aesthetics of concrete surfaces.

TEST PROGRAMME

The test programme consists to cast concrete specimens in moulds having different shapes and made with different materials. After demoulding, the surfaces were carefully observed to subjectively assess their aesthetics properties. For comparison purposes, all specimens were made with a self-levelling and an ordinary concrete mixture. Tests were carried out to assess the mechanical and the rheological properties of those mixtures. Microscopic examinations and ASTM C 672 deicer salt scaling resistance tests were also carried out on some samples.Although many different mould shapes had been used, most of the concrete specimens were cast in vertical flat moulds of relatively large dimensions. A flexible membrane was often placed on one side of the mould in order to produce some expressive forms, which are quite unusual for concrete (see Figure 1). Some objects (such as the maple leafs in Figure 4d) were sometimes glued on the formwork to produce stamped patterns at the surface of concrete. In some cases, complex shapes were chosen, or steel rebars were added, in order to obstruct the flowing of concrete. Ordinary concrete was always carefully cast in place using vibration and providing the best possible casting conditions. On the contrary, self-levelling concrete was put in place with no special care using neither vibration nor any other compaction mean.

PROPERTIES OF CONCRETE MIXTURES

The concrete mixtures were carefully designed to achieve optimal characteristics. Their composition and properties are given in Tables 1 and 2 respectively. Both mixtures were air entrained in order to be representative of the concretes commonly used in Canada for outdoor exposure conditions. The ordinary mixture was optimised in order to produce a particularly good reference concrete. The high-performance self-levelling mixture was made using a Class F fly ash, a napthalene based superplasticizer, and a viscosity admixture (wellan gum).

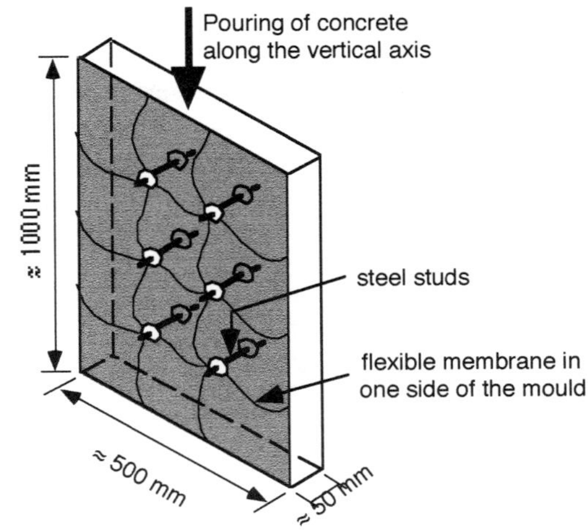

Figure 1 Typical example of formwork used in the study

Table 1 Composition of the concrete mixtures

	ORDINARY CONCRETE	SELF-LEVELLING CONCRETE
Water, kg/m^3	166	180
ASTM Type I Cement, kg/m^3	370	450
Silica Fume, kg/m^3	-	15
Fly Ash, kg/m^3	-	110
Fine Aggregate, kg/m^3	710	780
14 mm Coarse Aggregate, kg/m^3	1070	785
Superplasticizer, L/m^3	-	8,25
Water Reducer, L/m^3	0,81	0,45
Air-Entraining Agent, L/m^3	0,06	0,30
Viscosity Agent, kg/m^3	-	0,20

Table 2 Properties of the concrete mixtures

		ORDINARY CONCRETE	SELF-LEVELLING CONCRETE
Water to Binder Ratio		0,45	0,31
Unit Weight, kg/m^3		2320	2326
Slump, mm		115	260
Air Content, %	after 15 minutes	6,0	5,2
	after 45 minutes	4,6	5,0
Compressive Strength, MPa	at 3 days	37	46
	at 7 days	43	54
	at 28 days	50	68
	at 56 days	-	74
Flextural Strength, MPa	at 28 days	-	6,5

RHEOLOGICAL PROPERTIES

The rheological properties of concrete mixtures were recorded up to 1 hour after the initial contact between cement and water. The slump and the flow values were recorded at 15 and 45 minutes. A rheometer apparatus was also used to better assess the rheological behaviour of the mixtures at 30 and 60 minutes. This apparatus, described elsewhere [1] measures the force required to impose a circular motion to a special blade immersed in fresh concrete.

The results obtained are shown in Figure 2. In this Figure, the ordinate at origin represents the flow resistance i.e. the force required to put concrete in motion. The slope of the curve is called the torque viscosity because it is directly proportional to the plastic viscosity of the concrete mixture.

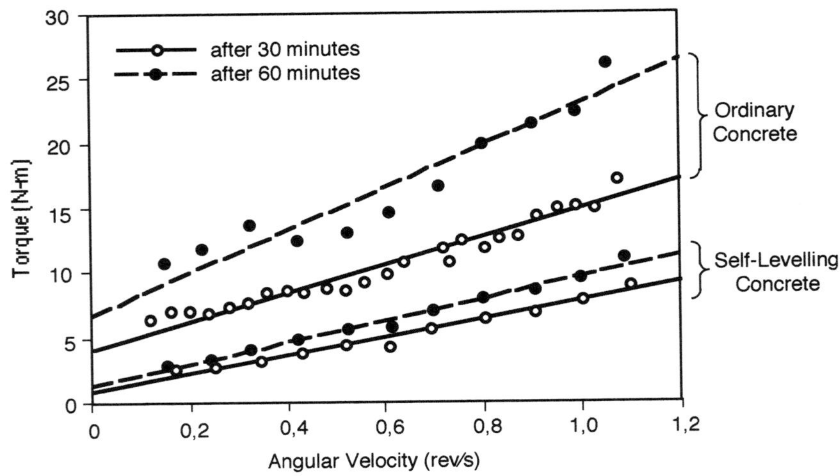

Figure 2 Test results obtained with the rheometer apparatus

Table 3 summarises the results obtained. One can notice a stiffening of the ordinary concrete since, over a 30 minute period (from 15 to 45 minutes after mixing), the slump value decreased from 115 to 75 mm. The flow resistance and the torque viscosity also increase by approximately 60% from 30 to 60 minutes after mixing. No such stiffening was observed for the self-levelling concrete. The slump remained the same (260 mm) and the flow decreased only slightly (from 600 to 500 mm) from 15 to 45 minutes after mixing. Within the 30 to 60 minute period, the torque viscosity was kept almost constant and, although the flow resistance increased significantly (from 0,9 to 1,5 Nm), it still remains quite low as compared with ordinary concrete.

It is worth noting that all the concrete specimens were cast not later than 45 minutes after mixing, which means that, at that time, the ordinary concrete had a workability within the limits commonly accepted at the job site (100 +/- 25 mm).

Table 3 Evolution of the rheological properties

	TIME AFTER MIXING, MINUTES	ORDINARY CONCRETE	SELF-LEVELLING CONCRETE
Slump, mm	15	115	260
	45	75	260
Flow, mm	15	-	600
	45	-	500
Flow Resistance, Nm	30	4,4	0,9
	60	7,2	1,5
Torque Viscosity, Nms	30	10,3	7,3
	60	16,2	8,1

AESTHETICS OF MOULDED SURFACES

As far as the aesthetics is concerned, the results obtained are very convincing and the tremendous superiority of self-levelling concrete is unquestionable. As compared with ordinary concretes, the self-levelling mixtures offer three major advantages:

1. They produce a surface with almost no air bubbles, compaction voids, or other surface defects;

2. They produce a very smooth and uniform surface with different appearances depending of the nature of the formwork;

3. They reproduce with a very great accuracy all the forms, textures, and printed details (even the thinnest ones) of the mould's surface.

For example, Figure 3 shows the surface obtained after demoulding when the self-levelling concrete mixture was simply poured into a formwork similar to that illustrated in Figure 1 without using vibration or any other compaction mean. The surface obtained has an exceptionally good-looking appearance with no voids or other surface defects, a uniform colour, a very smooth surface and a glossy finish. It can be seen that the concrete surface reproduces the texture and the folds of the plastic membrane very accurately.

Figure 4 shows other examples of the concrete surfaces obtained with self-levelling concretes. In Figure 4a and 4b, the concrete accurately reproduces the weaving pattern of the plastic membrane as well as the folds and the other very thin details of the formwork. In Figure 4c, the texture and the glossy finish of concrete both contribute to achieve a strong presence of concrete an architectural element. The Figure 4d shows the print left by a maple leaf on the concrete surface. All the details of the leaf were scrupulously reproduced, even the thinnest ones. It is also significant to notice that in all those Figures, no voids or other surface defects are visible. No specimens made with ordinary concrete had achieved such a high degree of visual perfection.

For instance, Figure 5 shows the photographs of two moulded surfaces obtained with ordinary concrete. It is clear that, even in the best conditions, ordinary concrete cannot satisfactory reproduce all the thin details of the formwork nor it can provide a surface as smooth and glossy than the surface obtained with self-levelling concrete Ordinary concrete also yields surfaces with a significant number of voids and other surface defects. Those voids are relatively large and they seriously compromise the good-looking appearance of the concrete surface.

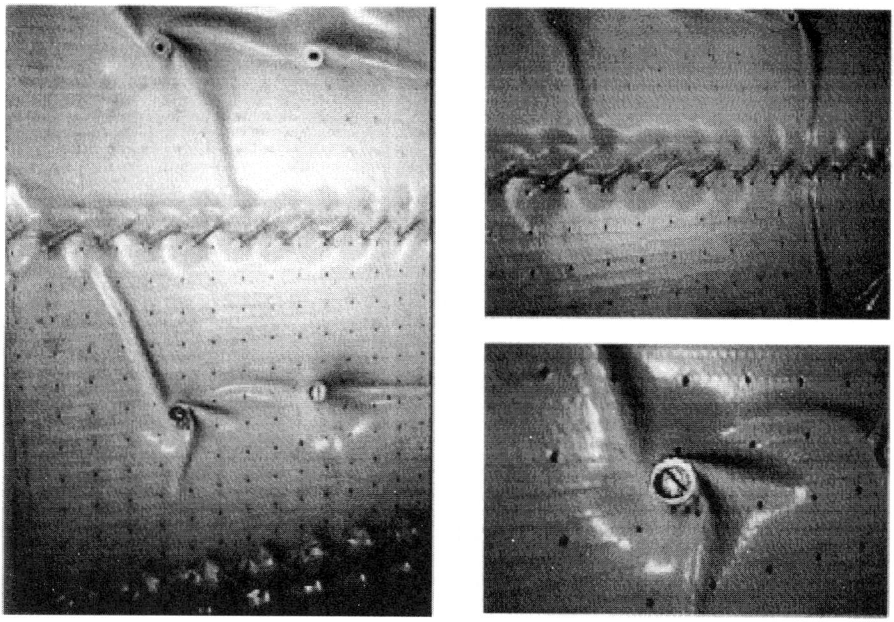

Figure 3 Photographs of a self-levelling concrete moulded surface

Figures 4 and 5 eloquently illustrate the huge potential of self-levelling concrete from an aesthetics point of view. This potential was confirmed by other laboratory tests made in our laboratory but not reported in this paper. The cast-in-place concrete thus acquires a genuine and incontestable aesthetic, or even artistic, value. It can be used not only as a structural material, but also as an attractive element which plays a significant role in the architectural design of the building.

A number of observations can be drawn from the laboratory tests. First, the texture of the concrete surface is highly dependent on the material used for the formwork. Porous materials, like wood or fabrics, produce a mat surface, which reflects light only slightly, if any. On the contrary, non-porous materials yield very glossy surfaces, which are highly reflective such as those shown in Figure 3. The glossiest finishes were obtained by using plastic formworks presumably because this material is non-porous and water-repellent.

Aluminium is also an interesting material in making the formwork because it chemically reacts with Portland cement to form gaseous hydrogen which produces bursting bubbles at the surface of concrete. This surface thus acquires a granular texture which is quite similar to the appearance of broken stone. Research actually carried out in our laboratory indicates that it is possible to obtain a wide range of surface finishes and textures by using different materials in making the formwork. For the architect, the choice of those materials thus become a matter of concern because it will significantly influences the visual aspect of concrete after demoulding.

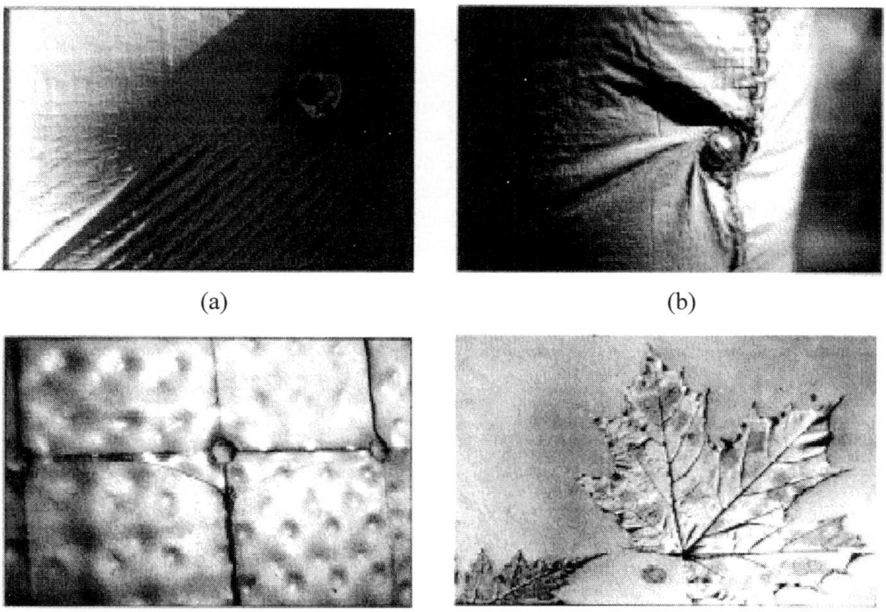

(a) (b)

(c) (d)

Figure 4 Photographs of self-levelling concrete surfaces

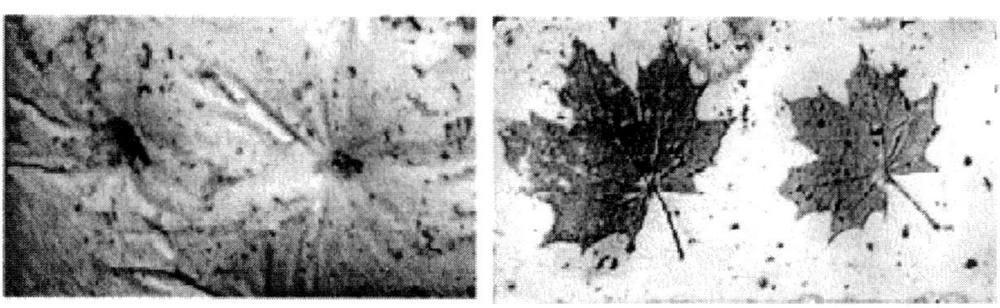

(a) (b)

Figure 5 Photographs of ordinary concrete surfaces

Test results also indicates that the compatibility between the cement, the superplasticizer, and the other constituents is a matter of concern (for some mixtures, the setting of cement was delayed up to 48 hours due to an incompatibility between the cement and the superplasticizer). The use of fly ashes was found to be particularly helpful to enhance the rheological behaviour of self-levelling concrete and to keep a good workability for a long period of time (up to 90 minutes). It was also found that, since self-levelling concrete flows almost like a liquid, a special care should be taken to ensure the watertightness of the formwork and prevent water leakage.

SURFACE PROPERTIES OF SELF-LEVELLING CONCRETE

In self-levelling concretes, the fluidity of the material is so high that the cement paste can flow easily throughout the skeleton of solid particles until it reaches the boundary of the mould. As a consequence, a thin layer of cement paste is thus formed near the surface of concrete. This surface layer is approximately 1 mm thick and is clearly visible to the naked eye on broken surfaces of self-levelling concretes. Examinations under a scanning electron microscope further reveal that this layer is essentially composed of dense calcium silicate hydrates with almost no sand particles or air voids. The aesthetics properties of self-levelling concretes are obviously directly related to the particular nature of the microstructure in this thin surface layer. No such layer was observed with ordinary concretes.

In some cases, map cracking was observed on the moulded surface of self-levelling concretes. This map cracking was barely visible to the naked eye and becomes really apparent only on wet surfaces. Hairline cracks are closely spaced (the distance between cracks is roughly equal to 1 cm) and they are strictly confined to the skin of concrete. No such map cracking was ever observed on ordinary concrete specimens.

The map cracking observed on self-levelling concretes is most probably due to the plastic shrinkage and/or to the drying shrinkage occurring in the cement paste forming the thin surface layer. It suggests that the curing conditions, especially in the early hours, are worth special attention in order to avoid surface cracking. According to the test results, map cracking is less likely to occur in self-levelling concretes containing fly ashes presumably because the lower hydration rate of fly ashes (as compared with Portland cement) tends to reduce early shrinkage.

The microstructure of the thin surface layer also raises concerns about the salt scaling resistance of the self-levelling concrete moulded surfaces. This resistance could be detrimentally influenced by the fact that the thin surface layer contains almost no air voids, has a low tensile strength and, most likely, a low permeability. In order to address this issue, four specimens of a self-levelling concrete mixture were subjected to the ASTM C 672 deicer salt scaling resistance test. The test was carried out on moulded surfaces very similar to those illustrated in Figure 3. During the test, two concrete specimens were covered with a brine solution (3% NaCl), and subjected to 50 daily freezing and thawing cycles. For comparison purposes, two other specimens were covered with pure water and subjected to the same freezing and thawing cycles. At different time intervals, the scaling residues were collected, dried, and weighted. The results obtained are illustrated in Figure 6, which shows the relationship between the mass of the scaling residues and the number of freezing and thawing cycles.

This Figure first indicates that the deterioration is negligible for scaling occurring in pure water, which is not surprising considering that even non air-entrained concretes are generally frost resistant in such conditions. It also indicates that, after 50 freezing and thawing cycles in the brine solution, the mass of the scaling residues is equal to 1,8 kg/m^2, which largely exceeds the usual 1 kg/m^2 acceptance limit. It is worth knowing that a loss of 1,8 kg/m^2 roughly corresponds to the destruction of 1 mm all over the surface, which is approximately equal to the thickness of the thin surface layer. It is not possible, however, to conclude that the poor scaling resistance is due to the presence of this thin surface layer. It is also possible that this poor scaling resistance is an intrinsic property of the concrete mixture tested regardless of the nature of the surface layer. This is likely, especially considering the fact that this mixture contained a significant amount of fly ashes, and that fly ashes are known to often have a detrimental influence on the salt scaling resistance of concrete [2]. With this respect, continuing the scaling tests up to 100 freezing and thawing cycles would be very interesting. A linear increase of the mass of the scaling residues up to 100 cycles would thus means that the poor resistance is an intrinsic property of the material. Otherwise the scaling process would slow down considerably after 50 cycles indicating that the poor resistance is mainly due to the presence of a weak surface layer.

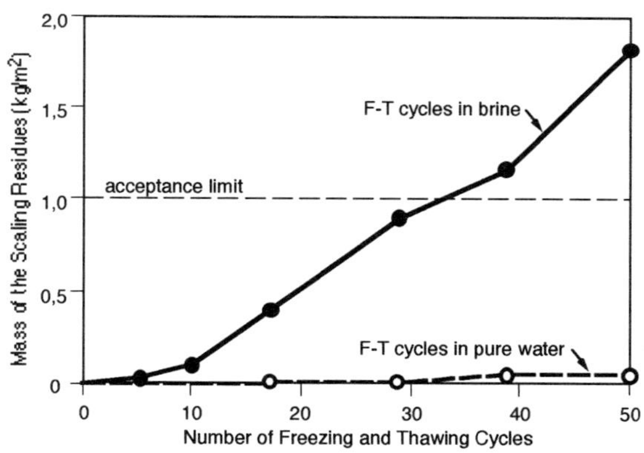

Figure 6 Deicer salt scaling resistance test results

CONCLUSIONS

The results reported in this paper clearly indicate that the recently developed self-levelling concretes offer a huge potential for enhancing the aesthetics properties of concrete. This can significantly contribute, not only to build concrete structures having a better looking appearance, but also to develop a wide range of new applications for concrete as a building material Self-levelling concretes also offer to architect a unique and genuine means of artistic expression which can not be achieved with any other material. Of course, research is still need to better understand and control the characteristics of these new concretes and to address some new technological issues (formwork's design, placing and curing techniques, durability, etc). This could be a stimulating challenge for the industry of concrete of the new millenium!

ACKNOWLEDGMENTS

The authors want to gratefully acknowledge the Laval University for its financial support to this research project. They also thanks Stéphane Allard, Vincent Augier, Stéphane Beal, Marc Drolet, Hélène Galifer, Alexia Gauthier, Mélanie Pinard, Emmanuelle Savarit, and David Umemoto for their help in "Creating with Concrete" at Laval University.

REFERENCES

1. BEAUPRE, D. Rheology of High-Performance Shotcrete, Ph.D. Thesis, Department of Civil Engineering, University of British Columbia, Vancouver, Canada, 1994, 250 pp.

2. PIGEON, M AND PLEAU, R. Durability of Concrete in Cold Climates, E & FN Spon Editor, London, England, 244 pp.

3. SEDRAN, T. Les betons autonivelants (BAN) — Synthese bibliographique, Bulletin de liaison du Laboratoire des Ponts et Chaussees, No. 196, 1995, pp 53-60.

CONCRETE IN SCULPTURE – THE IRISH EXPERIENCE

S MacCraith

University College Dublin

Eire

ABSTRACT. The extent to which concrete has been used in sculpture in Ireland is explored. The possibilities that concrete offers for this medium are examined and the problems experienced by sculptors in terms of materials, mixes, formwork, casting, curing and finishing are discussed.

Keywords: Concrete Sculptures, Public Sculpture, Curing, Vandalism, Moulds, Durability, Public Reaction.

Seoirse MacCraith is a retired lecturer in Civil Engineering, University College Dublin. At present he is visiting lecturer in Structural Design at the School of Architecture, UCD. Following graduation in Cork in 1953 he worked in the structural steel industry for six years. This was followed by 15 years in consulting engineering both with Veryard and Partners, Cardiff and Ove Arup & Partners, Dublin. He has spent the last 24 years in University College Dublin. In 1992 and 1995 he was Course Director for the Dublin Advanced Concrete Technology course of the Institute of Concrete Technology. Apart from his normal university duties he has written and lectured on concrete durability, design in reinforced concrete for practising engineers, the engineer and music, problems of conservation, etc.

INTRODUCTION

In Ireland, funding is available to local authorities from the Department of the Environment and Local government for the provision of artistic features under the "Percent for Art Scheme". This scheme, which is presently under review, allows a local authority, at its discretion, to provide an appropriate artistic feature in conjunction with and linked to a construction project being funded by the Department. Proposal from artists are invited by means of an open competition.

The origins of the scheme go back to 1978 when it was established by The Office of Public Works and in 1986 it became the "Per Cent for Art Scheme" of the DOE. Public sculpture of course goes back a long time before 1978 but there is since the scheme was introduced a growing awareness and appreciation by everybody of the role and importance of both private and public sculpture.

The Irish Concrete Society has been operating an awards scheme for concrete structures, including sculptures, for many years. However, both due to the increasing use of concrete in sculptures and the difficulty of judging the merits of a concrete structure versus a concrete sculpture, the Society decided in 1987 to introduced a separate awards scheme for concrete sculptures. This has increased the interest of sculptors in the potential of concrete as a medium for sculpture.

While all the traditional materials have featured in sculptures in Ireland, as elsewhere, there is now a considerable and growing use of concrete as a medium for small and large sculptures around the country, helped in no small way by the Government incentive.

The purpose of this paper is to highlight, by way of example and by allowing the sculptors to speak for themselves, the extent of this use of concrete for sculptures in Ireland. The problems of the material itself, the mixes, the casting and the finishing are all addressed. Eight concrete sculptures have been chosen and described in detail and a list of other sculptures is also given although the list is by no means exhaustive.

QUESTIONNAIRES

In order to assemble as much information as possible, a list of the sculptors who had entered for or had expressed an interest in the Irish Concrete Society's sculpture award scheme over the years was obtained from the Society. A total of 61 questionnaires was circulated to these and others and the results of eight of the 25 or so returned are included in this paper, and are shown in Figures 1-8. A list of other sculptors is given in Table 1. The selected sculptors' accounts best speak for themselves but, by way of summary of all the questionnaire returns, the following points emerge.

Wholly Concrete?

Ten of the sculptures were wholly concrete. The others included as additional materials mild steel, expanded metal, polystyrene, mosaics, wood, rubble, reeds, turf, ceramic, plastic, marble dust, bronze, glass fibre, glass, stained glass, water, a bicycle.

Size/Weight

There was a large range here. The lightest recorded was about 25 kg (*The Drinking Bird* – timber cased in concrete) and the heaviest was about 50 tonnes (*Perpetual Motion* – a 9 m diameter hollow sphere beside a motorway). The longest was about 250 m (*Telepath*) and the tallest the 9m diameter sphere.

Concrete Mix, Moulds and Finishes

Mixes were mostly 3:1 or 5:1 sand cement, some with admixtures (e.g. lime, fibres etc.) others without. Some included gypsum plaster. The amount of water tended to be judged to suit the particular sculpture and the mode of finishing. Concrete was mostly hand mixed but for larger sculptures ready mixed concrete was supplied. Many sculptures were formed around an internal armature but in some cases a clay cast was made from which a plaster mould was made and this then used to form the concrete shape. The nature of a particular sculpture dictated what the best approach was and no rules could really be formulated as to what was ideal.

The surface textures, specially important for a material such as concrete, were provided in various ways. Filing back and finishing with spatulas and trowels, sponging, use of gloved hands, sanding, finishing with lump resin, finish off the "wax carton" shutter, covering with ceramics or glass or mirrors, acid etching, pebble dashing, throwing water on the wet concrete followed by careful trowelling, cleaning and polishing with diamond pads and then waxing, polishing with beeswax and turpentine, modelling tools,- these were all used to provide suitable finishes.

Problems, Durability and Public Reaction

So far, durability is not a problem, since most of the sculptures are relatively young. A change in the colour of concrete over time has occurred – some darkening, some greening, some efflorescence. These, however, are not necessarily considered undesirable. Vandalism presents much more of a problem. Delicate parts of a sculpture, such as antlers and legs, have been broken rather easily, especially by children. In another case, *Maritime Piece*, in Wicklow, itinerants used the sculpture as clothes poles! Most sculptors have, through experience, learnt how to cope with this, but it is difficult to design for the serious vandalism such as the deliberate smashing of the *Goats* in a public park in Dalkey, Co. Dublin.

Public reaction ranged from acceptance to surprise to delight to astonishment – that concrete, which is often perceived as being ugly and vulgar, could be used in such an interesting way.

Observations of Sculptors

Nearly all of the sculptors who ventured an opinion on the use of concrete in their work expressed satisfaction with it. The following are some of the comments by sculptors recorded in the questionnaires, apart from those quoted in the examples given.

- "The method is relatively inexpensive in materials and equipment."
- "Since it has to be manmade, it has to be given more energy in order to exist as a visual substance."
- "Because of curing times, there is danger of vandalism between construction stages."
- "Excellent, but I do not use it enough."
- "I would love to work with it directly onto a frame or mesh."
- "A wonderful material and so economical."
- "Very pliable. Great for surface textures and ornament."
- "The fluidity of the material which turned into a solid form was exciting. Would like to work with the material in future."
- "Not easy to model with but possible. For me concrete sculptural work is largely experimental."
- "My experience is minimal. However, ----I am interested in its properties."
- "A medium that is not being exploited properly by the Architectural Profession, it's much more flexible."
- "As with all things, it depends on how it is used".
- "Interesting to start with a soft medium and end in a more permanent solid."
- "Student loved it. (It is) versatile, inexpensive, totally durable. Underused by sculptors."
- "For me it has great potential. Just getting into the medium really."

A Selection of Questionnaire Results

1. Name of sculptor
2. Location
3. Wholly concrete?
4. Other materials?
5. Size/weight
6. Concrete mix
7. Method of casting
8. Surface texture
9. Who manufactured conc.
10. Transport to site
11. Problems?
12. Durability, vandalism?
13. Public reaction
14. Observations on use of conc.

Group of School Children 1970

1. Imogen Stuart, R.H.A.
   Co. Dublin.
2. Tyrrell's Pass Green, Co. Westmeath.
3. Wholly concrete.
4. Aggregate of sand and fine stone
5. ?
6.
7. Firstly modelled in clay and mould made of plaster of Paris.
8. When cast (1970) the surface texture was particularly nice because of chemical reaction of plaster of Paris.
9. Home made.
10.
11.
12. So far no problems, except that one foot came off but was perfectly repaired.

13. Very popular. No more postcards of it available, but should be taken up by somebody!
14. Wonderful, but I am too old now for further work in concrete unless somebody else does it for me.

Figure 1 Group of School Children by Imogen Stuart

The Time of Day 1986

1. Cillian Rogers
   Co. Sligo
2. Easkey, Co. Sligo
3. Mainly
4. One bike + reinforcing steel + expanding mesh
5.
6. 1 cement : 2 sand (clean measured) + Evobond + frost proofer.
7. Hand moulded, section by section. Steel frame covered by expanding mesh..
8. Small trowels + modelling tools.
9. Ordinary hardware shop cement.
10. Car trailer.
11. This was one of my first concrete works. So it was an experiment.
12. The Community do not maintain it, as specified by me originally. It requires painting each year. It is sat and climbed on, every day and has survived fairly well considering.
13. It is photographed every day by tourists. It gets huge public reaction.
14. Concrete as a material has not been explored properly in Ireland. Its potential has not been exploited. Concrete technology is difficult to get information on in Ireland. I do

not think there are any sculptors in Ireland who really understand its true potential. I have a lot to learn yet.
(I have also made *The Black Pig* (Muc Dubh) in Inishcrone, Co. Sligo. Brian Laden wrote a novel based on me placing this sculpture in the town. The book - "Death and Plenty" is published by Brandon Press. Another piece is in Kiltamagh, Co. Mayo. Title: *Two Men and a Dog.*
I like to explore the comic potential of concrete.

Figure 2 The Time of Day by Cillian Rogers

Goats 1987

1. Katy Goodhue
   Co. Tipperary.
2. Dillons Park, Dalkey, Co. Dublin.
3. Mainly
4. Mild steel/expanded metal.
5. 1.5×life size.
6. 3:1 sand:cement + Black pigment. P.V.A added to mix.
7. Ferro concrete plastered onto armature.
8. Built up filed back and finished with spatulas and trowels to smooth finish.
9. Made at Symposium.
10. Dún Laoghaire County Council.
11. The standing goat had one leg fractured during siting, there were too many helpers and I was I think too inexperienced in people management to limit the number of people who lifted at once.
12. The horns were originally made of concrete on steel but this proved a bad combination as the steel had too much spring in it and the concrete was gradually fractured off it. The horns were replaced the following year with ones made of steel.

13. The pieces were made during the Tír Ciullain Symposium and the maquette was on show in Dalkey before the piece was sited.
14.

Figure 3 Goats by Katy Goodhue

Perpetual Motion 1994 – 1996

1. Rachel Joynt and Remco de Fouw.
   Dublin
2. Median strip between slip road to Naas and M7 Motorway at Maudlins, Naas, Co. Kildare.
3. Reinforced concrete shell.
4. Rebar and tubular sub-frame.
5. 9 m sphere = approx. 50 tonnes. Surface area is 250 m^2.
6. Fibre (polypropylene) reinforced shotcrete mix.
7. Spray applied to galvanised mesh on to tubular sub-frame. The whole work was fabricated by David Magee of Trench Control Ltd. Naas.
8. Trowelled hand float finish with sporadic 10 mm pebble hand thrown and pressed in with float to simulate bituminous road surface.
9. Goode Concrete Ltd., Naas, Co. Kildare.
10. Skeleton frame manufactured off-site at Trench Control Ltd., and delivered by truck. Shotcrete applied on site.
11. Curing and constructing a scaffolding to access the ball for painting. These problems were overcome. David Magee designed an ingenious rotating scaffolding as shown in photos.
12. Efflorescence – salt deposits and streaks. Otherwise durability is very good. We applied a resin paint called *Resutile*, a high grade two-part, light fast, chemically resistant fungicidal paint.

13. There was overall a positive public reaction and it got good press coverage.
14. Suitable for large pieces – requires early cooperation with sculptor and builder. To create a sculpture on this scale concrete is suitable and more feasible. This is particularly appropriate in the environs of a motorway where small works can appear 'lost'.

Figure 4 Perpetual Motion by Rachel Joynt and Remco de Fouw

The Deer of Scoil San Treasa 1996

1. Georgina Sutton
   Cork.
2. St. Teresa's School, Mount Merrion., Co. Dublin.
3. It is made from concrete around a steel frame.
4. Steel – steel wire. Ceramic.
5. 5.5' × 5.5'
6. Sand and cement (white) mixed.
7. Welded a steel frame (armature) put steel wire netting over this. Carefully layed on layers of cement.
8. Textured to a slightly stony effect. Using a technique of throwing water on wet cement. Careful trowelling.
9. Irish Cement.
10. Truck.
11. We were lucky with dry weather as we worked outdoors.
12. Children liked to hang out of antlers. We had to re-weld the antlers and use *Sika-dur* glue to lay the cement finely on the antlers.
13. Very good.

14. It is very exciting because you can achieve lovely shapes and combine other materials. We used ceramic for the eyes, nose and ears and feet of the deer.

Figure 5 The Deer of Scoil San Treasa

Concrete Embrace 1997

1. Jean Greene (Irish Concrete Society prizewinner 1997)
   Co. Dublin
2. Irish Cement, Stillorgan, Co. Dublin.
3. Wholly concrete
4. No.
5. 45 kg.
6. Aggregate from Armagh. Limestone + white cement + water.
7. Clay sculpture, plaster cast, clay removed, moult washed and cleaned.
   Concrete poured, waste moult removed.
8. Cleaned and polished with diamond pads and waxed.
9. I mixed the concrete.
10. By car.
11. Not really this time.
12.
13.
14. It is very hard work. But very rewarding when I get the finish I want.
    Concrete Embrace is mother with two children. I was very pleased with the colour – very warm with a nice fleck.

Figure 6 Concrete Embrace by Jean Greene

Jean Butler 1997

1. Des Gogarty
   Co. Meath.
2. At sculptor's address
3. Mainly concrete
4. Armature and netting of steel.
5. 6 ft. high, about 1 ton.
6. 1 to 5 silica sand.
7. In some sculptures I have cast the face, otherwise it is built direct.
8. Smoothed as concrete was setting and sanded afterwards.
9. Irish Cement.
10. Trolley.
11. No.
12. They have weathered very well.
13. Astonishment.
14. A wonderful material and so economical.

Liberation 1998

1. Holger Lönze
   Belfast
2. Belfast studio.
3. Wholly.
4. Structural armature mild steel.

Figure 7 Jean Butler by Des Gogarty

5. Height 165 cm, weight circa 80 kg.
6. (I used Ciment Fondu from builder's merchant).
   First layer: goo of 1 cement: 2 sand (building sand).
   Other mix: 2 cement: 6 sand: approx. 1 water. (All by volume)
7. Building in clay over armature, moulding (5 caps) in plaster/scrim.
   Then filled with goo layer; 2 – 3 layers of standard mix, inlays of fibreglass mats (chopped strand mat, rough, E-glass).
8. Plastermould – surface (very exact), polished with beeswax/turps.
9. Mixed by myself in studio.
10.
11. Warping of plaster mould – caps caused slight imprecision on joints.
12. Kept inside so far, cannot tell.
13. Generally, interest of concrete being used not for building or "ugly" structures.
14. Great stuff: cheap, easy to work with, amazing detail it picks up.
    Good durability.
    Use it for working with students: gives good understanding for other purposes as in architecture.

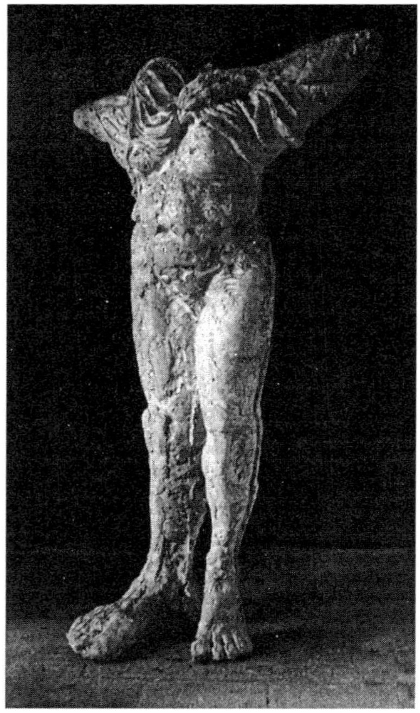

Figure 8 Liberation by Holger Lönze

Table 1 Other Irish Concrete Sculptures

TITLE	SCULPTOR	LOCATION
Water Feature	K. Goodhue	O'Reilly InstituteTCD
Secret Gatherer	K. Goodhue	Marlay Park
A Walk among Stone	*M. Bulfin	Ballymun
Aquaduct;	N. O'Neill	Donnybrook
Archer II;		Glasthule
Roller Coaster	N. Walsh	Killiney (private coll.)
Telling Tales	M. McGinty	Tallaght, Dublin
Sheep	V. Browne	Tallaght
Walk on me	A. Davey	Botanic Gdn.
Droichead na hAislinge	C. Reina	Ranelagh, Dublin 6
Reflection	A. Duncan	Dublin 16
The Ready Boat Pillar	S. O'Dwyer	West Pier, Howth
Cloud Star Boat Map	*G. Weir & F. Murray	Viking Centre, Dubl.
Power Icon	M.A. Gillan	Dublin
Fallen Figure.	B. Duggan	Pallas St'ios, Dublin 1

Table 1 contd. Other Irish Concrete Sculptures

Isambard Kingdom Brunel	J. Power	Sé Sí Original Shop, Temple Bar
Chiretsu Ní	S. Thomas	Dublin
Sculpture for Airport Roundabout	S. Mulcahy	Ranelagh (models only – never built)
Three Druids	C. McCormack	Marlay Park
Island;	R. Halpin	Guinness Dub
The Source		Drogheda
Collection	E. O'Connell	Celbridge
Ammonite	N. O Neill	Dublin Airport
Wall Sculpture	R. Costelloe	Ballsbridge, Dublin 4
Living Relic	*M. Warren	Dunleer Bypass
Pope John Paul; etc.	O. Grant	Enniscorthy
Installations 1 and *2;* etc.	M. McGinty	Donegal
Heraldic Door; Freedom	A. Hall	Valentia Island, Co. Kerry
Chapel Floor	J. Scanlon	Co. Limerick
Maritime Piece; Shepherd	V. Browne	Wicklow
		Tinahely
Waterfall of Souvenirs	J. Kindness	Bus Station, Belfast
The Thrones	M. Warren	Leighlinbr'ge
Drinking Bird; etc	M. Casey	Newt'cashel, Co. Longford
Male Bust	M. Wilkins	Co. Galway
Sand Serpent	J. Forrester	Playschool, Cork
Jean Butler; Ancient Irish Warrior; etc	D. Gogarty	Julianstown, Co. Meath.
Bualadh Bos	K. Herbert	Studio, Belfast
Telepath	J. Mallon	Glaslough, Monaghan
Facets of the Female Form	J. McKenna	Rathbride, Co. Kildare
Serpentine Seat	O. Kaminska	
Furling her Hoop	S.A. Morgan	Drogheda
Decorative Murals	R. Dobson	Ballymena and Belfast
Church and Sculptures	N. Pollen	Maghera,
	I. Stuart	Co. Tyrone
Murals	B. Mitchell	N. Ireland
108 heads (plaster)	Leah Durkan	Windy Gap, Co. Mayo
Mural		Youghal, Cork

* prizewinner

CONCLUSIONS

Concrete has proven itself as a worthwhile medium for sculptures in Ireland. Sculptors have enjoyed using it and have demonstrated in many ways the versatility of the material. Many of the sculptors are members of the Sculptors' Society of Ireland and has cooperated over the years with the Irish Concrete Society in promoting the use of concrete in sculpture.

ACKNOWLEDGEMENTS

The Author acknowledges with gratitude the help he has received from The Irish Concrete Society and from the Sculptors' Society of Ireland, in providing information and names for this paper. In particular, however, very sincere thanks must go to the various sculptors who returned the questionnaires or who otherwise provided the relevant information and who facilitated the taking of photographs.

The Author (Tel. **353 1 2895465) has full names and addresses and telephone numbers of the sculptors for any readers who are interested in seeking further information. The Sculptor's Society of Ireland may also be contacted at **353,1,8722364

TWO REMARKABLE ROUNDABOUTS IN REINFORCED CONCRETE REALISED IN FLANDERS (BELGIUM)

L Hendrikx

FEBELCEM

C Caestecker

F Venstermans

Flemish Brabant Roads Division

Belgium

ABSTRACT. In recent years, several roundabouts have been constructed in continuously reinforced concrete (CRCP), particularly in the Walloon Region of Belgium and mostly with a large diameter (35-80 m). On the initiative of the Flemish-Brabant Roads Division, the first Flemish roundabout of CRCP was successfully completed in 1998 in Tienen on the Ring R27. This roundabout with an internal diameter of 33 m was constructed in one pass with a Belgian DRION-slipform paver. In May 1998 a very special roundabout with an internal diameter of 14 m was constructed as part of the reorganisation of the centre of Scherpenheuvel, using decorative green concrete with exposed coloured aggregates.

Keywords: Roundabout, Concrete, Pavement, Continuously reinforced concrete (CRCP), Exposed aggregate, Colour, Design, Aesthetic

Ir Lambert Hendrikx, is Head Engineer – Infrastructure at the Federation of the Belgian Cement Industry (FEBELCEM). He is a civil engineer and has great experience in road construction. He was a member of the International Technical Programme Committee of the 8th International Symposium on Concrete Roads in Lisbon (1998).

Ir Chris Caestecker, is a civil engineer; he completed his studies in 1969. Since 1972 he was active at the Belgian Ministry of Public Works, where he was responsible for the construction of many important projects. Since 1991 he has been General-Inspector of the Flemish-Brabant Roads Division. He is also a member of the PIARC Committee C7 Concrete Roads.

Ir Frans Venstermans, is a civil engineer and since 1971 has been active in the construction and maintenance of roads in Flemish-Brabant.

INTRODUCTION

As in several other European countries, more and more roundabouts are being constructed at busy crossroads in Belgium. In order to prevent the road surface from being distorted by heavy loads and to prevent the top layer from moving as a result of horizontal forces, a pavement of concrete, slabs and if possible, continuously reinforced concrete (CRCP) is often recommended.

In recent years, several roundabouts have been constructed in continuously reinforced concrete, particularly in the Walloon Region of Belgium and mostly with a large diameter (35-80 m). On the initiative of the Flemish-Brabant Roads Division, the first Flemish roundabout of CRCP was successfully completed on 30 March 1998 in Tienen, on the Ring R27. This roundabout with an internal diameter of 33 m and a road width of 7 m was constructed in one pass with a Belgian DRION-slipform paver.

A few weeks after the construction of the extremely successful roundabout in Tienen the Flemish-Brabant Roads Division together with the town of Scherpenheuvel-Zichem came up with another first. In May 1998 a very special roundabout was constructed as part of the reorganisation of the centre of Scherpenheuvel, using decorative green concrete with coloured aggregates in the top layer of the pavement with reinforced slabs, to create a seven-pointed star.

The concrete was washed out; enhancing the appearance of the coloured granulates. This roundabout has an internal diameter of 14 m and a road width of 6 m. The concrete, 70 mm top layer and 180 mm underlayer, was placed between fixed forms. First, the green points of the star were made.

This paper describes the details concerning the design, the concrete mix and the construction of both remarkable roundabouts.

THE FIRST FLEMISH ROUNDABOUT IN CONTINUOUSLY REINFORCED CONCRETE IN TIENEN

The roundabout is located on the new section which connects the N9 to the N221, and in this way adds to the completion of the Ring R27 around Tienen.

Characteristics

The internal diameter of the roundabout is 33 m. The width of the road is 7 m (2 x 3,5 m).

The structure is as follows:
- 230 mm continuously reinforced concrete
- 40 mm bituminous layer (dense asphalt)
- 200 mm lean concrete
- 200 mm sand for foundation.

The longitudinal reinforcement consists of bars with a diameter of 16 mm. For good flexibility of the reinforcement, it was decided to use a diameter of 16 mm and place them at intervals of 120 mm, instead of the usual 150 mm. There is overlapping of 560 mm (Figure 1).

In this design, the percentage longitudinal reinforcement amounts to approximately 0,72 %.

Figure 1 View of the reinforcement

The transverse reinforcement supporting the longitudinal reinforcement is placed on supports with a height of 120 mm. The transverse reinforcement has a diameter of 14 mm and was placed at an angle of 60° to the longitudinal joint (see Figure 2). An additional transverse reinforcement 5 m long, with a diameter of 14 mm, was constructed for the right-hand lane. The intervals of the transverse reinforcement are:

- 0,40 m at the level of the external circumference, 0,34 at the level of the longitudinal joints,
- 0,56 m at the level of the internal circumference.

The additional transverse reinforcement was considered necessary for extra strength and to close any possible longitudinal cracks.

The reinforcement design was determined in consultation with FEBELCEM, the Research Centre of the Cement Industry and the Flemish-Brabant Roads Division. The contractor, Betonac (St.-Truiden) completed this reinforcement plan in a very thorough manner.

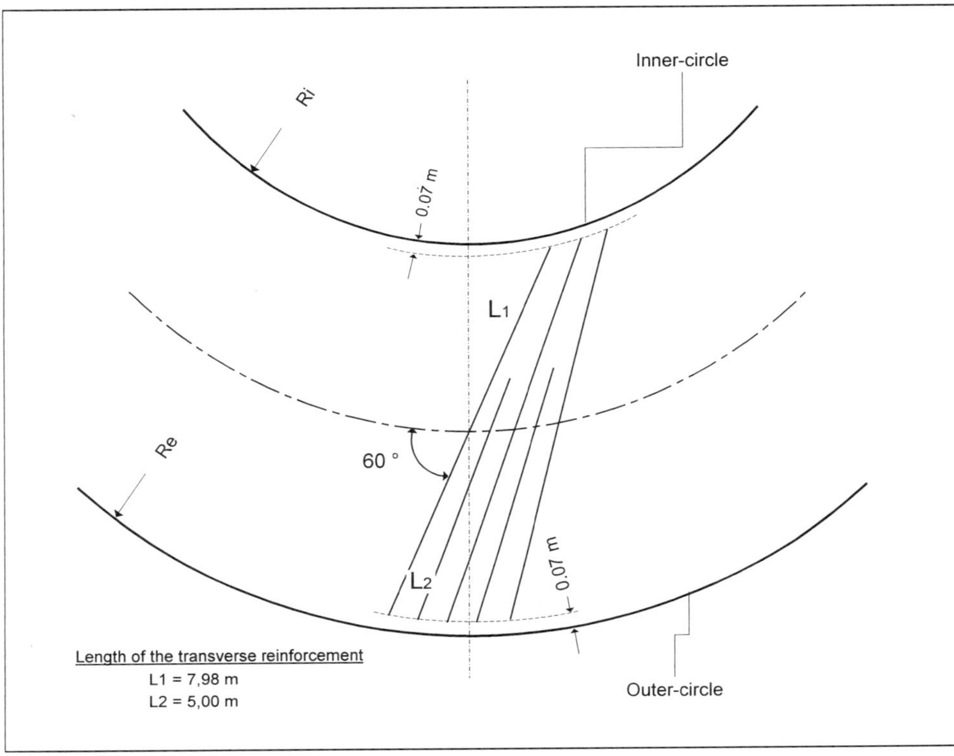

Figure 2 Detail transverse reinforcement

Construction

As the ring around Tienen was constructed in low noise concrete slabs, low noise concrete was also chosen for the roundabout (0/20 with minimum 20 % 4/7). For roundabouts, this is actually not really necessary because the speed is restricted. However, exposed concrete increases the grip, and enhances the aesthetic appearance. The mix design of the concrete was studied in the laboratory of the Research Centre of the Cement Industry and was composed as follows:

- sandstone divided over 3 fractions, 4/7 (413 kg), 7/14 (338 kg), 14/20 (337 kg)
- river sand (684 kg)
- cement CEM III/A 42,5 LA (400 kg)
- plasticizer P 200N (0,2 kg)
- air entraining agent (0,4 kg)
- water (174 l).

With this mix design, there was an air content of almost 4 %. The compressive strength in the laboratory reached an average of 43 N/mm² after seven days (cylinders – 113 mm diameter).

The Drion-Glijbouw Company made the whole roundabout with a Drion-paver (slipform paver) with a width of 7 m (Figure 3). Only a strip of about 1 m at the end of the work was compacted manually.

A longitudinal joint was sawn in the middle of the surface.

The concrete mix was produced in the main contractor Betonac's own batchplant. This mix plant was set up close to the job site and was also used to prepare the concrete to make the pavement of the ring which Betonac did it itself.

For the construction of the roundabout, technical assistance was provided by the Research Centre of the Cement Industry under the supervision of P. Van Audenhove, the manager of the Cement Industry "Technical Assistance".

Figure 3 General view of the finished roundabout in Tienen

A TWINKLING ROUNDABOUT WITH COLOURED EXPOSED AGGREGATES IN SCHERPENHEUVEL

A few weeks after the construction of the extremely successful roundabout in continuously reinforced concrete in Tienen, the Flemish-Brabant Roads Division together with the town council of Scherpenheuvel-Zichem, came up with another first. In May 1998, a very special roundabout was constructed as part of the reorganisation of the centre of Scherpenheuvel, using decorative green concrete in the pavement of the road, to create a seven-pointed star. The construction of this unusual project, designed by the Hagelands Survey and Study Bureau in Scherpenheuvel, was a real challenge for all those involved, and in particular for the contractor, Limasco.

Characteristics

The roundabout lies in Scherpenheuvel at the crossroads of the local roads (the Vesten and the Basilieklaan) and the main road N10, coming from Aarschot.

This recent reorganisation of the centre of Scherpenheuvel particularly concerns the renewal of the Noorder- and Zuidervest, making great use of concrete paving blocks (20.000 m^2) which are available in many shapes and colours, even green. For the roundabout, a hard-wearing monolith surface was opted for in reinforced concrete slabs, because of the heavy traffic causing torsion.

In order to give the concrete a more attractive and cheerful appearance, it was decided to use coloured concrete with exposed aggregates which have their own natural colour, such as grey and green. This technique, for which pigments were usually used as well, is aimed particularly at producing colour-fast surfaces.

By using the same colours as those in the adjacent materials, the designer achieved a perfect integration of the roundabout in the project as a whole, which is extremely successful because of the presence of some remarkable trees.

As the number seven has a symbolic significance in Scherpenheuvel - the basilica and the surrounding streets were built in the shape of a seven-sided figure, the designer aimed to represent a seven-pointed star in the surface of the roundabout. This was done by producing the seven points in concrete with green exposed aggregates (Figure 4).

The roundabout has an internal diameter of 14 m, and the width of the road is 6 m.

Figure 4 Form of the seven-pointed star

Taking into account the special configuration of the concrete surface and the relatively high pressure resulting from the torsion of the traffic, the following structure was decided upon in consultation with the numerous partners, including the town of Scherpenheuvel, the Flemish-Brabant Roads Division, FEBELCEM and the Research Centre of the Belgian Cement Industry:

- 70 mm concrete with exposed coloured aggregates (C 35/45)
- 180 mm bottom layer concrete C 30/37
- 40 mm dense asphalt
- 180 mm lean concrete
- 250 mm sand for foundation.

Figure 5 Detail of the reinforcement

Because of the special shape of the surface with numerous sharp corners, it was considered necessary to provide a heavy reinforcement mesh namely 150 x 150 x 12 mm (Figure 5). The reinforcement was interrupted at the level of the islands in order to construct doweled contraction joints.

Due to the small diameter of this roundabout and also because of the special configuration CRCP was not appropriated for this construction.

On the large sections of grey concrete there are contraction joints, 70-80 mm depth. A longitudinal joint also had to be sawn in the middle of the road.
In addition, it should be noted that in order to avoid the sharp corners on the edge of the surface, the points of the star were truncated.

Construction

The mix design of the concrete with exposed coloured aggregates is as follows:

– chippings (coloured)	980 kg
– river sand 0/3	720 kg
– cement CEM III/A 42,5 LA	425 kg
– admixtures (pigment - air entraining agent - plasticizer)	25 kg
– water (total)	± 180 l

With this mix composition an air content of slightly more than 3 % was achieved, which is normally adequate to make concrete with a very high cement content resistant against de-icing products. The compressive strength after seven days on drilled cores out plates 40 x 50 cm, made of this concrete on the job site, was well above 30 MPa. The compressive strength of the bottom layer was above 40 N/mm² after 7 days.

The mix design of the coloured concrete was determined by the Research Centre of the Belgian Cement Industry and is practical the same as the mix of low noise concrete (0/7). The contractor laid the concrete pavement in two layers, a bottom layer of 180 mm normal concrete and a top layer of 70 mm coloured concrete. All the concrete was placed between fixed forms. First, the green points of the star were constructed (Figure 6).

Figure 6 Completed point of the star

Despite the complicated design of the surface, the contractor was able to carry it out very satisfactorily.

In order to achieve a good join between the green and the grey concrete, the reinforcement continued between two joints.

The concrete was compacted with vibrating needles and a double vibrating beam. When it was completed, the Pieri Benelux firm sprayed a surface retarder VBA2000 (blue) free of solvents onto the newly laid concrete.

One day after spraying the retarder, the concrete was washed out enhancing the appearance of the coloured granulates.

A protector was successfully used to protect the natural stone of the central island, which was also partly constructed of washed out decorative grey concrete. This serves as an anti-joining layer and allows for the easy cleaning of the adjacent paving blocks.

Before starting on construction, several preparatory discussions took place with all the partners involved, under the leadership of the design engineer. This was certainly necessary for an efficient and satisfactory construction.

Figure 7 General view of the coloured roundabout

CONCLUSIONS AND RECOMMENDATIONS

1. Heavy traffic places great demands on the road surface of a roundabout.

2. Concrete pavements, both concrete slabs which could be reinforced and continuously reinforced concrete, are extremely suitable for this. If the internal diameter is greater than 20 m, it is now possible to use continuously reinforced concrete adequately with the help of a slipform paver.

3. Roundabouts with a diameter of less than 15 m can be constructed durable in reinforced concrete slabs. They must be constructed between fixed forms.

4. By using the two-layer system, it is possible to construct strong and extremely decorative concrete road surfaces in an economical and technically satisfactory way. Because of the fairly high pressure on the edge and the torsion of the traffic, a thickness of at least 230 mm of concrete is recommended for roundabouts on district roads.

5. In the case of CRCP the longitudinal and transversal reinforcement must be adapted.

6. A bituminous layer of 40-60 mm dense asphalt between the lean concrete base and the concrete pavement can certainly be recommended. In addition it is desirable to use a subbase of at least 180 mm of lean concrete.

7. The texture of the surface can be chosen in harmony with the environment without any problem. Pigments and coloured aggregates provide all sorts of possibilities (Figure 7).

8. Special projects require very thorough preparation and close collaboration between all the parties involved to achieve the best possible results.

INDEX OF AUTHORS

SUBJECT INDEX

This index has been compiled from the keywords assigned to the papers, edited and extended as appropriate. The page references are to the first page of the relevant paper.